Inheritance Theory: An Artificial Intelligence Approach

Raad Al-Asady
Exeter Unviersity

Printed in the United States of America

Library of Congress Cataloging-in-Publication Data

Al-Asady, Raad.
Inheritance theory : an artificial intelligence approach / Raad Al-Asady
p. cm.
Includes bibliographical references and index.
ISBN 1-56750-155-9 (cloth). -- ISBN 1-56750-156-7(paper)
1. Artificial intelligence. 2. Commonsense reasoning. 3. Logic, Symbolic and mathematical. I. Title.
Q338.85.A4 1995
006.3--dc20 95-44227
CIP

Ablex Publishing Corporation
355 Chestnut Street
Norwood, New Jersey 07648

Contents

This book is dedicated to my children
Zaid, Yasmin,
& Hala

in the hope that one day they
may follow what is behind it.

Foreword

Dr. Al-Asady's book tackles a deceptively simple-looking problem: commonsense reasoning such as we all do, effortlessly, all of our lives. The universality of the human ability to reason sensibly about everyday matters suggests that it might be an important, basic component of that special human quality - intelligence.

This is, in fact, the view taken in this book - commonsense reasoning is seen as a fundamental ingredient of intelligence. It is, moreover, for anyone who has not thought about it in detail, a surprisingly difficult phenomenon to reproduce within a formal framework such as the science of AI demands.

How do we manage to reason successfully without knowing all the facts - and knowing that we don't know them all? And when we have arrived at conclusions that, although reasonable at the time, later show themselves to be patently false, what do we do about it? We reject the false conclusion, but what about all the other lines of reasoning (and resultant conclusions) that may have been based on this acknowledged false conclusion?

Definitive answers to such questions are not yet available, but reasonable and well-defined answers must be found if we are ever to be able to add the crucial ingredient of commonsense reasoning to an AI system. In this book, you will find some answers to these sorts of questions about the basics of human reasoning.

Derek Partridge
Exeter University

Preface

This book is based on my doctoral dissertation, completed at Exeter University, in 1993. It is concerned with creating an inheritance theroy from the AI point of view. Within Artificial Intelligence, the need to create sophisticated intelligent behavior based on commonsense reasoning has long been recognized. Such commonsense reasoning is characterized by the withdrawing of previously drawn conclusions when new information comes along. Research has demonstrated that formalisms for dealing with commonsense reasoning require *nonmonotonic* capabilities where, typically, inferences based on incomplete knowledge need to be revised in the light of later information that fills in some of the gaps.

In this book, an *inheritance theory* based on multiple inheritance structures with exceptions (nonmonotonic inheritance structures) is proposed. Without an adequate nonmonotonic inheritance reasoning technique, such as *exceptional inheritance reasoning* (EIR) as proposed in this volume, inheritance networks will produce inconsistencies. A number of nonmonotonic properties that enable EIR to subsume existing formalisms such as default logic and inferential distance ordering have been included within the reasoning technique presented here.

An inheritance algorithm is also presented and a demonstration is included to show how it can be used to specify and implement various nonmonotonic inheritance problems. In addition, an inheritance formalism has been developed that is capable of dealing with ambiguous situations and can handle other classes of nonmonotonic problems apart from those already presented in the literature.

Finally, in the application section, this inheritance formalism has been applied to two important domains, namely causal reasoning and analogical reasoning, to demonstrate the conceptual power and expressiveness of the formalism.

Acknowledgments

Initially, and before anything, I thank and pray to God who gave me the belief, the health, and the strength to do this book.

I would like to thank Ajit Narayanan, Antony Galton, Derek Partridge, and Brian Lings, among other staff members and postgraduate students in the department of computer science at Exeter University, who have supported, discussed, and reviewed this research work with patience.

Also I would like to thank my parents, my wife Kamila, and my brothers and sisters for their love and support. Without them this research book would not have been possible.

Raad Al-Asady

Chapter 1

Introduction

1.1 What is Intelligence?

Although different definitions of AI have been suggested in the past, they all share the stipulation that an intelligent machine be capable of acting independently, of having a degree of self-determination. There are many aspects of intelligence; for example: learning, behavior changing, problem solving, language comprehension, and using reasoning. All these may be considered important features of intelligence. According to this definition, if an entity can make its own decision without any external help then we may call it an intelligent entity.

It may be that humans, in making decisions, use a kind of algorithm or procedure that they vary according to the specific situation. Decision making by natural entities could follow a right or a wrong algorithm. By and large, if a decision is based on sound reasons then we call it a *reasonable decision* or a decision that makes sense.

One particular area explored by this book is the exploitation of *commonsense knowledge.* The kind of reasoning that uses commonsense knowledge is called *commonsense reasoning.* Commonsense reasoning is a phrase referring to the intellectual abilities that are special to humans; in other words, intelligent entities make use of commonsense reasoning. Commonsense reasoning involves reasoning about objects, the objects' behavior in different situations, and the relationships that hold between objects. This type of reasoning is not characterized by a specific formalism such as logical deduction, yet it is commonly used by intelligent entities.

One aspect of commonsense reasoning involves the retraction of previous conclusions as new information comes along.

There is an implicit assumption that concept formalization can help an intelligent system to find solutions to problems. Formalizing concepts using mathematical and logical reasoning may let machines do things in a way similar to or better than humans. Unfortunately, it is difficult to construct a general formalization of commonsense reasoning. As a starting point, classical logic may be considered as a first attempt that provides procedures, usually using transitive deductive reasoning to formalize commonsense reasoning. However, the shortcomings of classical logic when dealing with exceptions and changes over time were addressed by the appearance of different modified versions of classical logic, designed specifically for this kind of reasoning which is called *nonmonotonic reasoning.*

In order to create plausible intelligent systems that can reason about the real world, a wide range of reasoning styles has been proposed. The technical problem with these proposals lies in their inability to describe human commonsense reasoning precisely or intuitively. The real world is dynamic, with new states of affairs arising out of old. Such new states replace old states of affairs, and new states must be represented in such a way that old states are cancelled out. *Monotonic reasoning* styles assume an incremental growth of knowledge, where new states and situations are continually added to the existing states and situations. But the real world does not operate in this manner. Any reasoning style that fails to take into account the simple fact that a new state may cancel out (i.e., contradict) a previous state (e.g., a robot starting at a door and moving to the other side of the room will no longer be near the door at some point in its journey) will be severely limited in its ability to capture real-world information or represent the dynamic nature of reality. On the contrary, almost all work in AI has focused on monotonic systems (with the result that the AI community as a whole has a rather poor understanding of what nonmonotonic systems are capable of). One of the dominant paradigms in AI is the use of logic-based reasoning styles, where for instance the derivation of a new fact is like deriving a theorem. In standard logic reasoning styles, there is no such thing as generating a possible fact and then checking its truth. To derive a possible fact is to derive a theorem, given the truth of the premises. Any such theorems can be thrown back into the melting pot of existing facts, and further theorems can be derived with the added theorems. There must be no contradiction between later theorems and earlier theorems and premises in a monotonic system. Any reasoning style that attempts to evaluate new facts in terms of their effect on existing facts, so that statements that are no longer true can be retracted to leave a consistent system state, is a nonmonotonic style system.

Many attempts have focused on formalizing commonsense reasoning (activities). Reiter's default logic (Reiter, 1980) and McCarthy's circumscription (McCarthy, 1980), among others, are examples of such formalizations.

1.2 Hierarchical Organization of Knowledge

On the one hand, the problem is that a human cannot always formally define a general procedure or an algorithm to follow in solving a problem in a specific situation. The shortcomings of humans may create the need for a more intelligent entity to solve a problem.

On the other hand, using an intelligent machine, we may discover a new paradigm that humans alone failed to find. For example, it has been suggested that the limitations of our natural languages express the limitations of our thinking. If this is true, and such limitations were to be overcome by machines, the machine may be able to achieve things we could not. Such an intelligent machine, if created, would change the way people think.

Intelligence needs knowledge. In other words, machines could do things better if they use something similar to a human's commonsense reasoning ability. But humans use knowledge about domains and past experience in solving problems; a machine with no experience or knowledge would have difficulty achieving some conclusions a human could achieve easily. This problem has led researchers to think about the role of knowledge in AI; that is if we need an intelligence machine we have to formalize human common sense reasoning and use human knowledge in different domains. A human's knowledge may be gathered in a machine by asking an expert to enter his knowledge into the machine. This is what we call building a knowledge base for an expert system[1].

In this section we shed some light on the importance of knowledge in general AI and demonstrate the different knowledge representational languages.

1.2.1 Representation of Knowledge (Facts) Using Logic

To illustrate, earlier in the literature, logicians such as Hayes, (Hayes, 1977), defined logic as a collection of ideas on how to express certain kinds of knowledge about certain kinds of world. There are three general ways in which logic can be used to model nonmonotonic reasoning. The first is to say that the way knowledge is represented in the system should be in classical, first order logic (i.e., using standardly formed predicates and arguments). Generally, classical first order logic provides procedures, using usual transitive deductive reasoning, where no exception is allowed. The problem with classical logic is that once we reach a conclusion, by valid reasoning, we cannot then retract this conclusion without also retracting at least one of the premises on which it is based. Because exceptions are permitted in reality, classical first order logic fails to supply the suitable mechanism. It is important to stress here that the issue has to do with whether general statements of the form "X's are Y's" can be correctly represented by logical formulae of the form $\forall X(X(x) \rightarrow Y(x))$, with the quantifier interpreted classically. The point is that in everyday thinking "X's are Y's" usually does not mean this. It means that X's are typically Y's, with the probability of an indefinite number of exceptions that cannot be specified completely in advance. Under such circumstances, we say that we have *nonmonotonicity*. Classical logic (FOPC) cannot cope with nonmonotonicity because, as mentioned before, it uses rules of universal validity that can be applied to facts to deduce unretractable conclusions.

Knowledge structures can benefit from all logic-based representation

[1]What normally happens is that a 'knowledge engineer' elicits the knowledge from the 'expert' and has to work hard to get it into a form that can be entered into the computer.

languages [2] (by utilizing their inference mechanisms) to 'jump over gaps' in our beliefs and knowledge. Let us demonstrate this by using one type of the aforementioned languages, say, default logic (Reiter, 1980).

The general form of default rules is:

$$\frac{\alpha(x) : \beta(x)}{\gamma(x)}$$

where α, β, γ are first order formula whose free variable is x, α is prerequisite, β is justification, and γ is consequent. The interpretation of this rule is: For any individual x, if $\alpha(x)$ is inferable and $\neg\beta(x)$ is not inferable, then $\gamma(x)$ can be inferred.

For instance;

$$\frac{elephant(Clyde) : grey(Clyde)}{grey(Clyde)}$$

could be read as: if 'Clyde is an elephant' is given and 'Clyde is grey' is consistent with what else is known, then 'Clyde is grey' can be inferred.

1.2.2 Representation of Knowledge Using Semantic Networks

Although the various types of logic are considered to be powerful forms of inference, it is not the only form that is useful for representing knowledge. Graphical representation, including frames and semantic networks, using *nodes* and *arrows*, is considered another powerful type of representation.

Generally, a hierarchically organized semantic network may be considered as a labeled directed graph of three levels (see Figure 1.1). At the lowest level we find nodes that represent instances and individuals such as Katy, Clyde, or John, and so on. At a higher (second) level (or levels) we find nodes that represent the concepts of entities, attributes, or states. We refer to these nodes by classes or types such as Cat, Tiger, Human, Car, and so on. The individuals at the lowest level are related to the classes at the second level by arcs or conceptual relation called *is-a*. At the top level, properties are attached to the second level by either the structural link *(is-a)* or by its negation as in 'Katy is-a cat' and 'mammal is-a living-thing'.

[2]Those languages are the subject of Chapter 3.

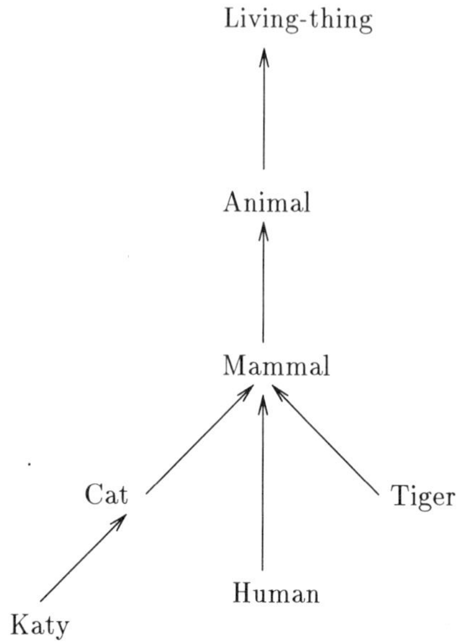

Figure 1.1: Inheritance of properties.

In order for a network, as a representation language, to be more effective and powerful, it is possible to augment the representation structure by an inference mechanism. One of the main inference mechanisms used within such language is *inheritance of properties.* Inheritance of properties arranges, in a linear way, each individual to its related property. According to this and Figure 1.1 we may deduce 'Katy is-a living-thing', (explained in some detail in the next section).

1.3 Inheritance

Inheritance is a mechanism for locating knowledge properly within a hierarchy in such a way that a specific property can be efficiently assigned to a specific entity. In addition, inheritance has the ability to assign all the relevant properties to a class so that these properties can be easily shared among different classes located within the hierarchy. Inheritance is considered a main mechanism for deducing information about more specific concepts from more general concepts with an efficient reduction in the storage requirements.

The motivation behind treating inheritance as form of path-finding lay in the hope that a variety of efficient graph traversal algorithms already existing in graph theory could be used for implementing the specialization and inference rules required in inheritance systems. For instance, early inheritance systems such as FRL (Touretzky, 1984) and NETL (Fahlman, 1979) used a form of shortest path strategy to implement such rules. Shortest path algorithms use either bounded depth-first or breadth-first searches and are provably correct for simple inheritance structures without exceptions. The problem arises when they are applied to nonmonotonic inheritance structures. This is the subject of Chapter 3.

Inheritance networks are based on *nodes* that represent individuals and classes with associated properties, and the *links* that denote the different relations between the nodes. The primitive relations are the *is-a* link and the *not-isa* link. The sole inference mechanism in these formalisms is inheritance between the hierarchical relationships.

Property inheritance may be explained by the following example (see Figure 1.2): if

Clyde *is-a* elephant,
elephant *is-a* gray.thing,
elephant is-a mammal,

then Clyde would act as if it were a copy of the entire structure at elephant. We could conclude that

Clyde *is-a* mammal,
Clyde *is-a* gray.thing.

In addition to the property inheritance itself we may assign the property value to an instance as well. For example, in addition to the previous example, the value of Clyde's legs is 4 because a typical elephant has 4 legs. In other words, Clyde, being an elephant, will inherit all typical properties of elephants.

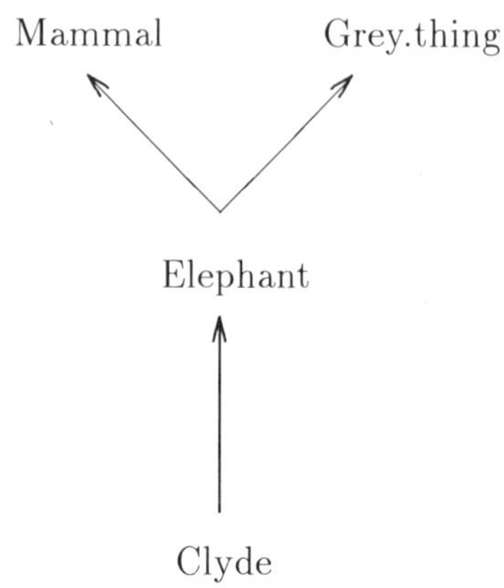

Figure 1.2: The Clyde-elephant network.

Inheritance is considered a main feature of semantic networks, and *inheritance networks* is used as another name for this type of representation of knowledge.

Although the normal use of inheritance is inheritance of properties down a hierarchical relationship, inheritance is usually associated with hierarchical representation formalisms or systems.

There is no relation between inheritance mechanism and the hierarchical formalization. To clarify, whereas an *is-a* link is used as a connection between nodes within a hierarchy, an inheritance mechanism is used as a vehicle to transfer knowledge across those links.

Previous attempts to represent nonmonotonic inheritance as a representation language (Fahlman, 1979; Touretzky, 1986) failed to capture exceptions in a general formalization language. Consider the following example (see Figure 1.3):

Clyde *is-a* African.elephant
Clyde *is-a* Royal.elephant
Royal.elephant *is-a* elephant
African.elephant *is-a* elephant
Elephant *is-a* grey.thing

Royal.elephant *not-isa* grey.thing.

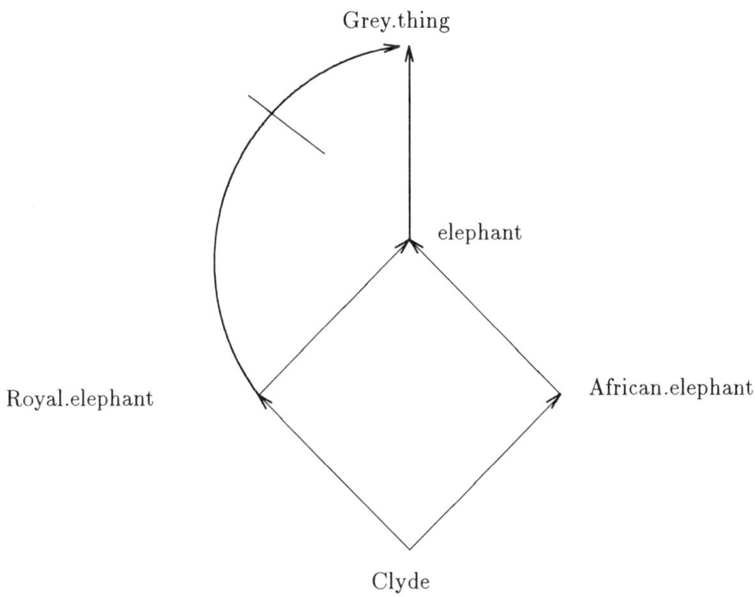

Figure 1.3: The Royal-African.elephant network.

In order to identify Clyde's properties, NETL (Fahlman, 1979), for example, chooses one of the two paths from Clyde to grey.thing/not grey.thing arbitrarily; TMOIS (Touretzky, 1984) failed to choose any one of them using a specific inference mechanism. Because no clear semantics is present within these systems, problems with inheritance as a weak representation language were clearly evident.

In this book I propose an inheritance theory, with a well-defined semantics, to represent and handle nonmonotonic inheritance structures. The representation schema is called *Exceptional Inheritance Reasoning* (EIR). The mechanism we use presents a new view of inheritance hierarchies. In the current literature, the *is-a* links are used to specify a superclass relationship as well as to infer inheritance relationships, a *not-isa* link specifies a non-superclass relationship but nevertheless requires a property to be inherited; namely, the opposite of what would be typically inherited. This is questionable: If *is-a* is used to specify class relationships as well as inheritance, then *not-isa* should be used for explicitly ruling out certain class relationships as well as not permitting any inheritance of properties of any sort. Exceptional inheritance theory has a clear semantics, a clear formalization, and an easy representation language.

In summary; representing a large amount of information about real domains is essential to AI. Generally, real domains are imprecise and contain uncertain and incomplete knowledge. Using an inaccurate representation leads to difficulties concerning maintainability and expressiveness. For these reasons a reasoning system must be able to manage and handle such domains through an appropriate knowledge representation formula for commonsense reasoning.

To overcome the shortcomings of knowledge representation, different formalizations have been presented and different semantics proposed, providing algorithms and procedures for identifying acceptable formula as well as translating networks into some standard logical formulae.

The subject of this book is an investigation into the nonmonotonic nature of inheritance networks. The problem is how to describe structures of classes and properties, with exceptions, in the right way; that is, based on a well-defined semantics that also gives rise to commonsense reasoning behavior. The difficulty here concerns the nature of the knowledge used, that is, the way this knowledge can be identified. In other words, this is defining what we mean by the *typicality* or the *normality* of classes, as well as the exceptions to this relationship between them.

The subsequent question is, what is the importance of defining the typicality or the normality of a class? Consider the following real example: I was invited to a restaurant. I could imagine what a typical restaurant means. However this restaurant happened to be a Japanese restaurant, where I did not expect to have to take my shoes off, which is beyond my definition or understanding of the typicality of restaurant[3].

Another example: If we are told that John is a typical PhD student then we will understand, among other things (using human commonsense

[3]It will be impossible for an artificially intelligent machine to understand what is beyond 'taking shoes off' without a clear understanding of typical restaurant.

reasoning), that John is a full-time student. But if John is a part-time student then there is an error either in our commonsense reasoning or in the definition of typicality. Because we cannot define the meaning of the typicality of a class unless we go through all the characteristics of that class, no precise single label will be enough to explain what we mean by a typical PhD student.

The problem with revising an old conclusion is that we could retract our original (true) statement in the light of a new one that might be true as well. The question is which one of them is true? Is it the old or the new? If it is the new, how does the truth of the new statement affect existing statements, including those derived from the original, no longer true, statements? There is an obvious need for an *algorithm* or a *procedure* to enable us to be able to indicate what we mean by *later* or *earlier* typical information.

In this book, Chapter 2 provides a review of the literature relating to inheritance networks. In that chapter I start by defining and describing the role of knowledge representation in Artificial Intelligence, then I illustrate the concepts of networks as well as inheritance. I focus on the mechanisms of multiple inheritance with exceptions and the problems that appear with such structures, namely redundancy and ambiguity. In addition I describe previous treatments of these problems.

Chapter 3 presents a survey of current approaches to nonmonotonic reasoning and its formalisms. I highlight the role of logic-based approaches, path-based approaches, and lattice-based approaches involving inheritance structures.

Chapter 4, summarizes the problem of handling conflicts generated by multiple inheritance structures involving exceptions, namely a clash of intuitions.

In Chapter 5, I present a new formalism for commonsense reasoning based on exception-based inheritance reasoning, namely revising an old decision with a new one without retracting the old one.

In Chapter 6, I tackle another inheritance problem, namely making decisions with conflicting extensions and ambiguous classes.

In Chapter 7, an application-oriented section, I apply exceptional inheritance reasoning to causal reasoning. In that chapter I define the importance of inheritance as a basis for defining causality and its reasoning.

Analogical reasoning is considered to be an important aspect of intelligent behavior. I use exceptional inheritance reasoning and causation as a basis for defining and applying analogical reasoning in Chapter 8.

Chapter 2

Inheritance Hierarchies

2.1 AI, Knowledge Representation and Inheritance

As mentioned in Chapter 1, the general problem of Artificial Intelligence (AI) lies in defining the right commonsense procedure that could assist in building an intelligent machine. This procedure needs to have the ability to simulate human commonsense reasoning to perform human tasks. One of the questions that arises is what the componentshum human commonsense knowledge or intelligence are.

From an AI point of view, human intelligence has three parts: facts and knowledge, a general method or rules for applying this knowledge, and a reasoning strategy that expresses the way these rules could act in a specific task or activity. We first start with some basic terminology.

Knowledge representation means encoding justified true beliefs into suitable data structures. It is concerned with the development of suitable notations for representing knowledge (Thayse, 1988).

A *hierarchy* is a semantic representation that places emphasis on the arrangement of entities in a clean organization of the represented world. In addition, a hierarchy accommodates the relationship between those entities in a proper way. The notion of hierarchy allows objects to be organized into classes so that instances of the same class have the same properties and more specific classes may inherit the properties of more general classes.

Inheritance is a mechanism for locating knowledge properly within a hierarchy in such a way that a specific property can be efficiently assigned to a specific entity. In addition, inheritance has the ability to assign all the relevant properties to a class so that these properties can be easily shared among different classes located along the hierarchy. Inheritance is considered as a main mechanism of deducing information about more specific concepts from more general concepts with an efficient reduction in the storage requirements.

Reasoning takes place when the inheritance mechanism tries to decide under contradictory situations *the relation* between an individual and a specific class of properties.

Following the semantic network concept, first adopted by Quillian (Quillian, 1968), an *inheritance hierarchy* may be considered as an ordered structure of knowledge. The nodes of such a hierarchy represent knowledge and the arcs represent relations between these nodes. Nodes at the top of a hierarchy have less specific information than nodes at the bottom. Almost all structured knowledge representation languages and object-oriented programming languages are designed around this mechanism. NETL (Fahlman, 1979), TMOIS (Touretzky, 1984), and FRL (Robert, 1977) among others are systems based on inheritance mechanisms. (These systems are explained in more detail in Chapter 3.)

2.2 Inheritance Hierarchy Components Represented as Semantic Networks

In this section I demonstrate the basic concepts of inheritance structures, such as classes, individuals, instances, and properties. Also I illustrate the types of relation between these concepts, namely inheritance structure links, such as *strict is-a* links, and *defeasible is-a* links, and the inheritance mechanisms behind them.

Generally, we may consider an inheritance hierarchy as a semantic network of three levels consisting of nodes and arcs (see Figure 2.1). At the lowest level we find nodes that represent instances and individuals such as Tweety, Clyde, or John and so on. There is no way in such a hierarchy to represent a relation, if any, *between* entities at this level. At a higher (second) level we find nodes that represent the concepts of entities, attributes, or states. We refer to these nodes by classes or types

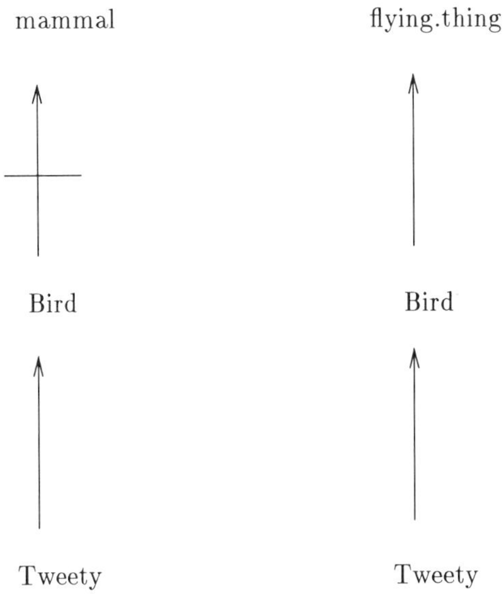

Figure 2.1: Tweety, the flying.thing.

such as Bird, Elephant, Human, Car, and so on. The individuals at the lowest level are related to the classes at the second level by arcs. An arc or conceptual relation (hereinafter *relation*) in a net represents a kind of structural link that holds between nodes. These arcs are represented in a net by the usual *is-a* link. At this level we can represent the relations, if any, between the individuals in the first level that share the same properties. At the top level, however, in order for the individuals to obtain a class' feature, the third level, namely classes of properties (hereinafter *properties*), is attached to the second level by either the structural link *(is-a)* or by its negation *(not-isa)*; (the different kinds of *is-a* links are explained later in this chapter) as in 'Tweety is-a bird' and 'Bird is-a flying.thing'. Also, we may represent the negation of *is-a* relation as in 'Bird not-isa mammal'.

The three inheritance structure levels may be augmented hierarchically according to the size of the represented knowledge (number of classes within one structure). Such expansion applies to the classes in the middle level. The classes that contain general information will be considered to be abstractions or main classes (hereinafter *superclasses*), and the more specific information will be considered to be secondary classes (hereinafter *subclasses*). For instance:

Clyde is-a Royal.elephant,

Royal.elephant is-a Elephant,

Elephant is-a Grey.thing,

Therefore Clyde is-a Grey.thing.

In the above relations, *elephant* is considered to be the superclass and *royal.elephant* is considered to be the subclass, because it contains more specific information than *elephant* regarding the individual Clyde. These relations may be augmented by attaching more subclasses to the superclass elephant, such as the subclass *African.elephant* or more specifically to the class *royal.elephant*, such as *king.royal.elephant.* Although these classes inherit their typical features from their superclass, they have some specific properties that could be different from the main superclass. Normally, the specific classes have the default properties of the superclass unless different properties are represented explicitly. This is explained in detail later in the book.

2.3 Inheritance Hierarchical Structures

To describe inheritance mechanisms, first we need to describe the different kinds of inheritance structures, namely, single inheritance structures and multiple inheritance structures. Later I will demonstrate the different kinds of problems that appear with these structures.

There are two types of inheritance structure:-

1. Tree structure or taxonomy hierarchy.
2. Multiple inheritance structure or directed graph structure.

We may consider a class as a collection of properties shared by instances and individuals, that is, if a knowledge base represents many instances and individuals sharing the same property, we might be able to create a joint class (superclass) that represents these instances.

Most KR's are organized around such an inheritance mechanism. This is because, in a hierarchy, using downward reasoning, properties can be imagined as features flowing downward from classes to individuals.

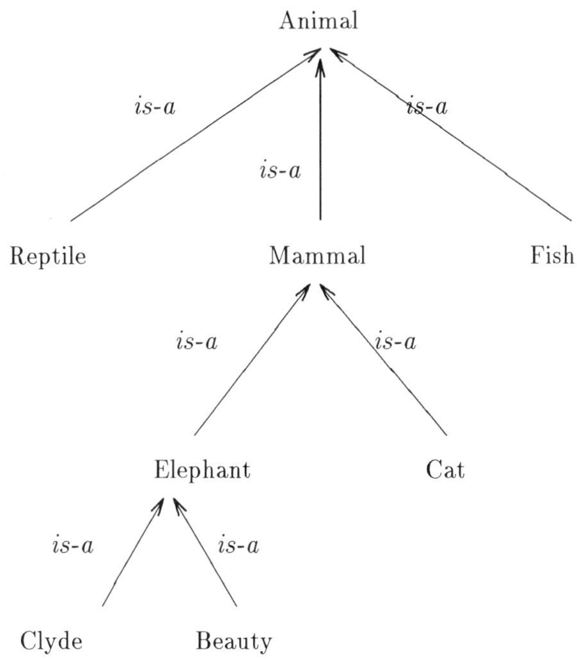

Figure 2.2: Tree structure.

A *tree structure* is a taxonomy[1]organized by the structural link is-a. For instance, we may read some of the relations represented in Figure 2.2 in the following way; *Clyde is-a Elephant* or *Cat is-a Mammal.* In order

[1]Taxonomy is a Greek word which means an arrangement based on any kind of law or principle (Sowa, 1992).

to obtain a new relation or a new extension (not explicitly represented in the network), deductive transitive reasoning is the main reasoning style adopted with such structures, such as *Clyde is-a Animal* (because Clyde is a elephant and a normal elephant is a mammal). In this kind of structure, each node has only one outgoing link. The number of inference paths in a structure can be determined purely through the number of individual nodes and the type of individual connections in the structure. The structural arrow points away from the leaves to the root nodes to specify generality. A property can be attached to an individual according to this procedure.

Classical first-order logic can describe such hierarchic taxonomies easily. For example, using Figure 2.2:

$$(\forall x)(Cat(x) \supset Mammal(x))$$

$$(\forall x)(Mammal(x) \supset Animal(x))...etc.$$

However, classical first-order representations could be precluded when exceptions are permitted with hierarchies. For instance, if we are told that a cat called Lucy *not-isa* Mammal (it happens to be a statue of a cat), our previous conclusion must be retracted, otherwise we will get two contradictory conclusions, namely Lucy *is-a* mammal, and Lucy *not-isa* mammal.

The shortcoming of tree structures[2] (as well as classical first-order logic) is a lack of ability in dealing with real-world knowledge. They do not have the ability to inherit a specific property from multiple ancestors as well as deal with exceptional information. In other words, a reasoning technique based on multiple inheritance structures with exceptions is needed with such structures.

As an alternative, multiple inheritance structures (or *tangled hierarchies;* (Fahlman, 1979)) were introduced. A multiple inheritance structure is an acyclic inheritance graph that can deal with the mentioned shortcomings. For example (see Figure 2.3):

Clyde is-a royal.elephant,

Royal.elephant is-a elephant,

[2] At this point we may distinguish between the double role of tree structure as a representation language and as an inheritance reasoning mechanism.

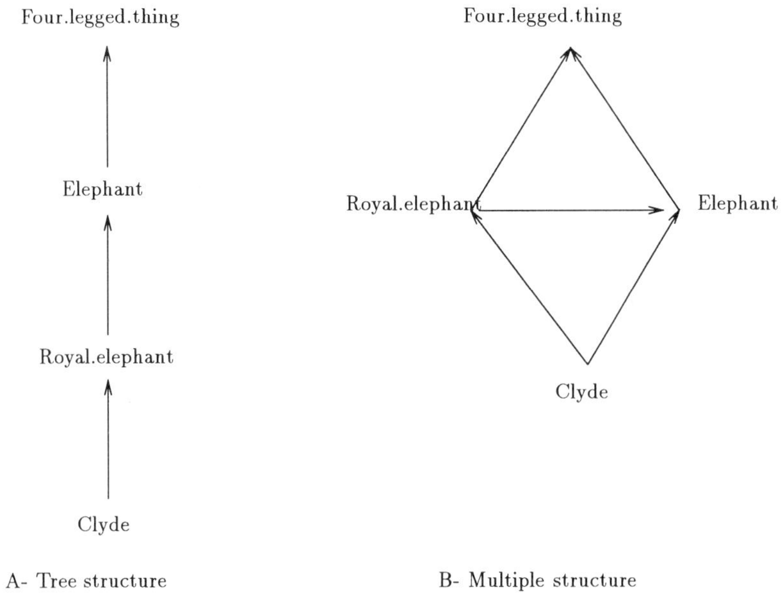

Figure 2.3: Clyde, the four.legged.thing.

Elephant is-a four.legged.thing,

Therefore,
Clyde the royal.elephant, the elephant, is-a four.legged.thing.

This example demonstrates the notion of multiple inheritance, for example, Clyde, the individual, acquires general properties through more than one class.

In real life, although individuals may inherit properties from more than one class, classes may vary according to their properties. Conflicts between these properties regarding one specific instance may occur.

For example (see Figure 2.4), an instance (A) may inherit a property (P) via a related class (C) and also inherit the negation of that property ($\neg P$) via another related class (D) which may also be related to class (C). The problem that arises then is from which of the classes will (A) inherit its property? If no special preferences or consideration

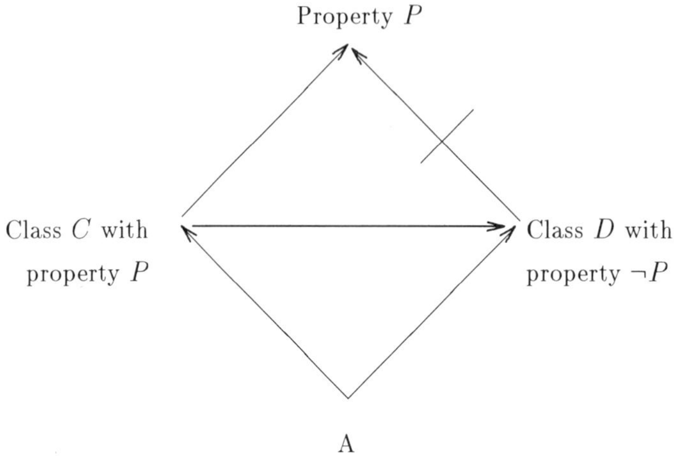

Figure 2.4: Contradictory classes.

exist between classes in their relations to an individual, all the conflicting properties will have an equally likely potential of being inherited by the same individual. That means we may assign the properties (P) or ($\neg P$) or neither or both randomly to that individual, which creates inconsistency. Because using such a mechanism, an instance may inherit a property and its negation via different directions, the number of inference paths in these graphs will be exponential to the number of their nodes.

Before proceeding to exceptions, a brief introduction to multiple inheritance notations and definitions is presented here.

An *inheritance hierarchy* or *inheritance network* is a labeled directed acyclic graph with nodes and arrows. Intuitively, nodes in a net stand for *individuals*, *classes* of properties or generic concepts (Touretzky, 1986), and *properties*. *Positive and negative links* (*is-a*, *not-isa*) are presented by the following arrows ($\rightarrow$, $\not\rightarrow$), respectively. *Is-a* link could be read as 'an instance of' or 'a kind of'[3]. An *is-a* link between an individual and a class stands for a positive relation such as 'Tweety is-a bird' or 'Clyde is-a elephant', and a *not-isa* link stands for a negative relation such as 'Tweety not-isa elephant'. Symbolically, we present the positive relation between an A and a class X as $A.X$, and the negative relation with a class Y as $\neg A.Y$[4]. A positive (is-a) relation between two classes (of properties) or between a class and its properties such as $x \rightarrow y$ stands for the statement 'an x is normally y' or 'an x is a subclass of y'. Similarly, a negative relation (not-isa) between two classes such as $x \not\rightarrow y$ stands for the statement 'an x is normally not a y' or 'an x is not a subclass of y'. There is no clear logical representation to such relation. Whereas a monotonic (strict) inheritance graph does not permit exceptions to inherited properties, a nonmonotonic (defeasible) inheritance graph permits exceptions, that is, a relation between two classes (of properties) can be explained as a defeasible relation because such a relation is subject to exception.

A sequence of related links is called a *path* or *relation*. Because an inheritance relation may include positive and negative links, a semantic inheritance structure will include *positive paths* and *negative paths* as well. All the paths in a network are *positive paths* unless the last link in a path is a negative link, in which case we call this path a *negative path*. A negative link terminates the path.

The *polarity* of a path represents the sign (positive or negative) of the final link in a path. For instance, if a path in a network includes the links (c_1, c_2), (c_2, c_3), ..., (c_{n-1}, c_n) then $c_1 \ldots c_n$ $(n \geq 1)$ is *a positive path*. Similarly, $c_1 \ldots c_{n-1} \neg c_n$ is a *negative path*. The nodes c_1 and c_n are called the *start point* and the *end point* of the path. An *intermediate node* in a path refers to any node located between the start and end points of that path.

The link formed by relating the start point and the end point of a path, together with the polarity of the last link of that path is called *conclusion* or *inference supported* by the path. A conclusion cannot be an intermediate link. An *inference rule* is a mechanism used to derive

[3]For more details about the different interpretation of the structural link *is-a* see (Brachman, 1983).

[4]This representation is not generally accepted. See, for example, (Thomason et al., 1987) in Chapter 3.

new paths from existing ones. An *extension* in a network is a path (or a set of paths) of form

$$x_1 \rightarrow \ldots \rightarrow x_n \qquad \text{or} \qquad x_1 \rightarrow \ldots \not\rightarrow x_n$$

supported by an inference rule. For example the path (A, c_1), (c_1, c_2), (c_2, c_3), ..., (c_{n-1}, c_n), (c_n, X) supports the conclusion 'an A is-a X'. Similarly the path (A, c_1), (c_1, c_2), (c_2, c_3), ..., (c_{n-1}, c_n), $\neg(c_n, X)$ supports the conclusion 'an A *not-isa* X'.

A class (of properties) is *inheritable* by an individual if all the paths from the individual to the class are positive paths, otherwise a class (of properties) is *un-inheritable. Un-inheritable* features may be caused by either *redundant links* or *contradictory classes.* These problems are discussed later in this chapter.

2.4 Exceptions

Within an inheritance structure, an *exception* is the negation of an inheritable structural link in a network; that is, negative links are used to override positive inherited properties, and positive links are used to override negative inherited properties.

Allowing exceptions of properties between nodes, two kinds of inheritance structures are recognized:

1.Structures containing *related* classes. Whereas some of these classes are related to one specific property, others are related to the negation of that property.

2.Structures containing *unrelated* classes. Some of them are related to one specific property whereas others are related to the negation of that property. No relation does exist between these classes.

Dealing with Point 1, the problem that follows is which of the multiple routes an individual will follow in choosing a specific property: the general class (superclass) via the positive link with a specific property, or the subclass via the exceptional relation to the same property?

From this, we can conclude that within inheritance structures there is no general concept that could hold true to all its related individuals and instances. Alternatively, to deal with real-life domains *a reasoning technique based on multiple inheritance structures with exceptions* is needed to provide a clean inheritance representation and support real-life nonmonotonicity.

In order to represent real-world domains, it is possible either to augment the standard inheritance structures (the single inheritance mechanism without exception) by exceptional relations, if any, or replace the existing mechanism with a more appropriate mechanism. We start by examining pre-emption.

For the remainder of this section we deal with the problem of exceptional inheritance structural links. The two problems that arise when exceptions are permitted in a hierarchy are the appearance of a redundant link and conflict classes.

2.4.1 Redundant Links

The redundancy problem (first adopted in (Touretzky, 1986)) is why explicitly linking two nodes previously implicitly linked by individual links should have side effects and affect what was previously derived.

Let us demonstrate the redundancy problem by an example. Consider Figure 2.5.

Tweety is-a Penguin,

Penguins are Birds,

Birds are flying.thing,

Penguins are not flying.thing,

Because the subclass Penguin has more specific information than the superclass, Bird, and because the instance, Tweety, reaches the property, flying.thing, via the class Penguin in fewer links (two links) than the class Bird (three links), then according to NETL (Fahlman, 1979) and TMOIS (Touretzky, 1986)[5] we reach the conclusion that Tweety cannot fly.

Now assume that later a user adds *explicitly* the following assertion (which was linked *implicitly*; see Figure 2.6):

Tweety is-a Bird.

This statement creates a problem (specifically to NETL, from (Fahlman, 1979)) that may be described as follows: While the link from Tweety reached the property, *flying.thing*, via the class bird (two links),

[5]These systems are explained in detail in Chapters 3 and 4.

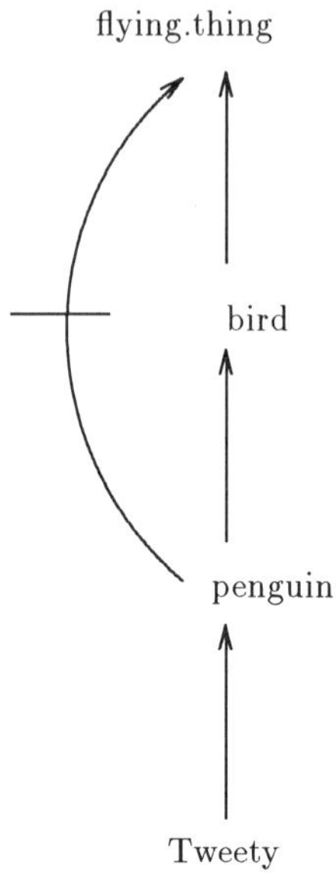

Figure 2.5: Tweety, the non.flying.bird.

other link from Tweety reaches the property, $\neg flying.thing$, via the class Penguin with the same number of links (two). No path (or marker in NETL[6]) could reach its property 'flying.thing' in fewer links than the other. In spite of the fact that the link $Tweety \rightarrow bird$ adds nothing to the reasoning mechanism regarding the property flying.thing, it creates a serious conflict problem. That is because this new path presents a (redundant) link that we cannot ignore or delete because it represents a completely true statement. Such redundant links in an inheritance structure will create uncertainty where previously there was none. In order to deal with such uncertainty, a procedure or an algorithm for preferring one of the two classes is needed. Such a procedure or algorithm

[6]NETL chooses the marker propagated by a class to its superclass that reaches its property 'flying.thing' earlier; that is, the shorter length among different paths using 'the computing marker propagation paths procedure' (Fahlman, 1979).

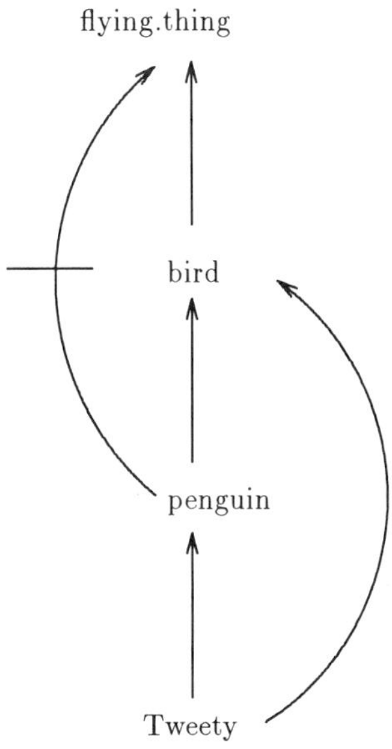

Figure 2.6: Tweety, the penguin.bird.

should also return the intuitively correct result.

In the literature, different approaches to the redundancy problem are proposed. For instance, Boutilier, (Boutilier, 1989) defined stable reasoning. A reasoner does not possess stability if the addition of explicit information to information already implicitly derivable has side effects in that it makes what was previously derivable, underivable. Boutilier claimed that calling a link redundant may give the impression that such a link adds no information to a network which is not the intention. Rather than calling these links redundant, *independently justified* may be a more appropriate term (Boutilier, 1989). In general, we may consider a link redundant in a network if there exists an alternative set of links that could be interpreted as the reason for the truth of that link. In the following chapter I demonstrate other approaches to this problem.

2.4.2 Ambiguity

Whereas redundancy deals with the procedure of choosing one class among *related* classes (in preferring a specific property to an instance), ambiguity deals with the procedure of choosing a class among *unrelated* classes. Most early inheritance systems, such as NETL (Fahlman, 1979), TINA (Touretzky, 1984), and FRL (Robert, 1977)) cannot even recognize ambiguity within their own mechanisms.

In the literature, and until recently, neither an algorithm nor plausible reasoning is adopted when a new conflicting fact is added to a knowledge base. The difficulty is due to the fact that both of the unrelated extensions correspond to a possible world state, and because no extra information is given in preferring one class on the top of the other, no reasoner could decide the actual, or even a reasonable, possible world-state.

Let us demonstrate with the Nixon diamond problem (Reiter, 1980) (see Figure 2.7).

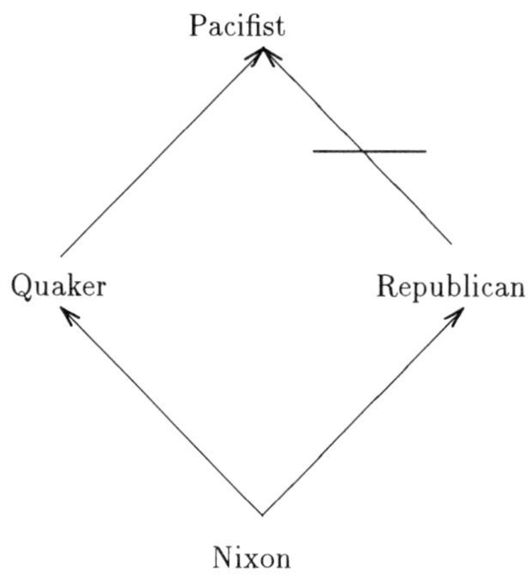

Figure 2.7: Nixon, the Quaker-Republican.

- Republicans are not pacifist.

- Quakers are pacifist.
- Nixon is-a Republican.
- Nixon is-a Quaker.

In this structure, because no relation exists between the two classes (Republican and Quaker), and because no extra information is available in preferring one class over the other, NETL (Fahlman, 1979) supports *either* the following two, as true extensions at the same time:

$$Nixon \rightarrow pacifist$$

and

$$Nixon \not\rightarrow pacifist,$$

or chooses one of them arbitrarily. TMOIS (Touretzky, 1986), on the other hand, generates the two as true extensions at the same time. The results of both these systems can be questioned.

2.5 Mechanisms with Inheritance Structures

As defined earlier, an inheritance hierarchy is a labeled directed graph whose nodes represent classes and individuals, and whose links denote various relations between these nodes. The motive behind such graphical (path-based) representations lies in defining a new extension (or conclusion) that is produced from a network using a reasoning style such as nonmonotonic inheritance reasoning. A variety of mechanisms have been proposed to handle path-based reasoning systems, including upward and downward concatenation, skeptical and credulous reasoning, and on-path and off-path pre-emption. This section demonstrates these mechanisms.

Concatenations: There are two directions that could be used to construct paths (path forming) out of primitive links in a directed graph, namely downward concatenation and upward concatenation. Downward concatenation can be described as an intuitive model of inheritance that

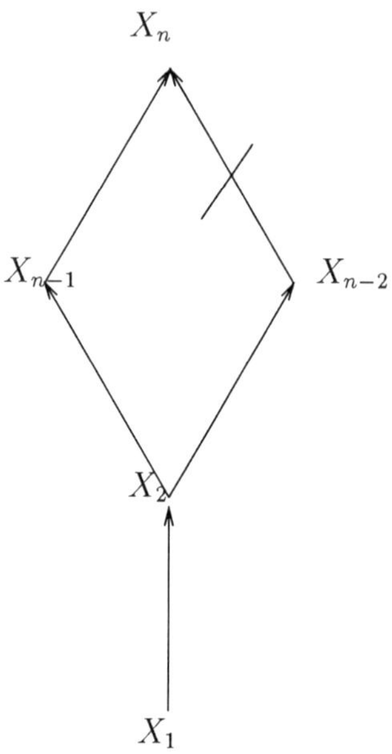

Figure 2.8: An inheritance network Γ.

has properties flowing downward from superclasses to their subclasses and individuals (Touretzky et al., 1987).

Downward concatenation can be defined as follows:

Definition 1 *Downward concatenation: An inheritance structure, such as* Γ *in Figure 2.8, with the sequence of links*

$$x_1 \rightarrow x_2 \rightarrow \ldots \rightarrow x_n$$

will permit either the path

$$x_1 \rightarrow x_n, \qquad \textit{or the path} \qquad x_1 \not\rightarrow x_n$$

only if the path

$$x_2 \rightarrow x_n, \qquad \textit{or the path} \qquad x_2 \not\rightarrow x_n$$

is permitted, respectively.

On the other hand, *upward concatenation* (Horty et al., 1987) may be described as follows:

> It is often natural to understand a path - like a proof - as representing a particular chain of reasoning behind the assertion it enables. The path $a \rightarrow p \rightarrow q$, for example, might enable the assertion 'Tweety flies', while representing an argument like 'Tweety flies because he is a bird and birds fly'. (Horty, et al., 1987, p. 359)

This approach seems natural when one wants to push the analogy between paths and arguments, because arguments tend to move from the beginning forward. From the given information, we may consider upward concatenation as an argument construction approach comparable with the downward concatenation approach (Horty et al., 1987).

Upward concatenation can be defined as follows:

Definition 2 *Upward concatenation: An inheritance network, such as* Γ *in Figure 2.8, with the sequence of links*

$$x_1 \rightarrow x_2 \rightarrow \ldots \rightarrow x_n$$

will permit either the path

$$x_1 \rightarrow x_n, \qquad \textit{or the path} \qquad x_1 \not\rightarrow x_n$$

only if the path

$$x_1 \rightarrow x_{n-1}, \qquad \textit{or the path} \qquad x_1 \rightarrow x_{n-2}$$

is permitted, respectively.

Credulous Versus Skeptical Reasoning: When deciding on a specific (preferred) extension, the literature is divided between two approaches. The first is called *credulous reasoning* (Touretzky et al., 1987). This approach supports as many extensions supplied by a network as possible (in order not to lose any possible extensions) so that inconsistency conclusions can be isolated from one another within their own extensions.

Credulous reasoning can be defined as follows:

Definition 3 *Credulous reasoning: An inheritance network, such as* Γ *in Figure 2.8, with the sequence of links*

$$x_1 \rightarrow x_2 \rightarrow \ldots \rightarrow x_n$$

and

$$x_1 \rightarrow x_2 \rightarrow \ldots \not\rightarrow x_n$$

will permit the path

$$x_1 \rightarrow x_n$$

and the path

$$x_1 \not\rightarrow x_n$$

in order not to lose any possible extension from the net Γ *(Touretzky et al., 1987).*

The second main approach is called *skeptical reasoning* (Touretzky et al., 1987). A skeptical reasoning system supports only one extension not containing any conflicting information, and refuses to draw any conclusion in an ambiguous or conflict situation.

Definition 4 *Skeptical reasoning: An inheritance network, such as* Γ *in Figure 2.8, with the sequence of links*

$$x_1 \rightarrow x_2 \rightarrow \ldots \rightarrow x_n$$

and

$$x_1 \rightarrow x_2 \rightarrow \ldots \not\rightarrow x_n$$

will support neither the path

$$x_1 \rightarrow x_n$$

nor the path

$$x_1 \not\rightarrow x_n$$

That is, it refuses to draw conclusions in conflict situations.

Within these approaches, the literature accommodates two streams: *ambiguity blocking inheritance* and *ambiguity propagation inheritance* (Stein, 1989). I demonstrate some systems that use these approaches in detail in Chapter 3.

Pre-emption: On-path Versus Off-path: *Pre-emption* embodies the basic idea of inheritance reasoning that, when two superclasses suggest contradictory properties, preference should be given to the property associated with the most specific superclass. We say that a path is pre-empted by another if the latter contains more specific information than the former. Consider Figure 2.9, where the net Γ supports the two conflicting inference paths, namely, $X_1 \rightarrow X_n$ and $X_1 \not\rightarrow X_n$. However, according to pre-emption X_1 will inherit the property $(\neg X_n)$ through the link $X_2 \not\rightarrow X_n$, rather than the property (X_n) through the path $X_2 \rightarrow X_3 \rightarrow X_n$, because the node (X_2) contains more specific information than the node (X_3).

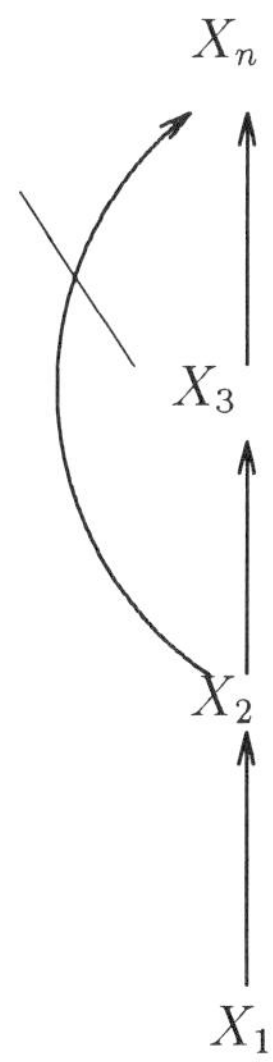

Figure 2.9: Γ: Pre-emption concept

There are two kinds of pre-emption. The first kind is *on-path pre-emption*, or inferential distance ordering, first described in (Touretzky, 1986). On-path pre-emption allows pre-emption to occur only if the pre-empted path contains a *redundant link.* The prerequisite for (on-path) pre-emption is that there are two paths of opposite polarity starting from the same node leading to the same end point. However, because one of these paths contains a redundant link, information associated with it (which is less specific and does not contain an intermediate node) will be overridden by information associated with the other (which is more specific and contains an intermediate node) *on* the path from the starting point to the end point. Formally, we may define on-path pre-emption as follows:

Definition 5 *On-path pre-emption: One path may pre-empt another only if the preempted path contains a redundant link that would short circuit part of the pre-emptor (Touretzky et al., 1987).*

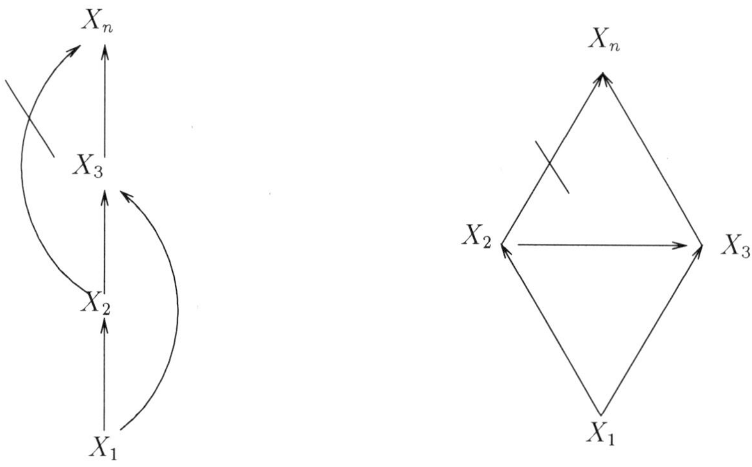

Figure 2.10: Γ1: On-path pre-emption.

For example, in Figure 2.10, the two inference paths $X_1 \rightarrow X_n$ and $X_1 \not\rightarrow X_n$ are supported by the net Γ1. However, because of the presence of a redundant link

$$x_1 \rightarrow x_3$$

within the path

$$x_1 \rightarrow x_3 \rightarrow x_n,$$

the path

$$x_1 \rightarrow x_2 \not\rightarrow x_n$$

will pre-empt the former because the latter contains (X_2) which is an intermediate node on the path, whereas the node (X_3) is not. Obviously, because on-path pre-emption's definition depends mainly on the existence of redundant links, on-path pre-emption will be criticized where there is no redundant link but conflict problems nevertheless exist. In order to prefer one path among others, of opposite polarity, and in the absence of a redundant link, *off-path pre-emption* (Sandewall, 1986) chooses the path that presents and adds *more explicit* information than the others. For example, if the following information is obtained from the net represented in Figure 2.11:

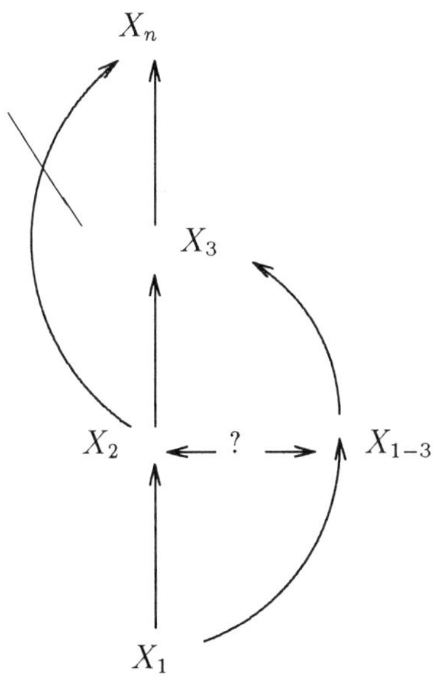

Figure 2.11: Off-path pre-emption.

1. The redundant link $X_1 \rightarrow X_3$ is interrupted by a new node X_{1-3}.

2. No relation does exist between the nodes X_2 and X_{1-3} (represented as $\leftarrow\ \ ?\ \ \rightarrow$).
3. Whereas the class X_2 has a negative relation with the end point X_n , X_{1-3} has a positive relation with X_n, because it is an intermediate node on the path $X_1 \rightarrow X_n$.
4. Because X_{1-3} location in the network is similar to the location of the node X_2 in that both of them are below the node X_3, the node X_3 becomes a superclass for both X_2 and X_{1-3}.

Under these circumstances, and because no redundant link exists, the link

$$x_2 \not\rightarrow x_n$$

overrides the inheritance path

$$x_1 \rightarrow x_{1-3} \rightarrow x_3 \rightarrow x_n,$$

because the former adds more explicit information to the network, whereas the latter contains just the (implicit) inheritance information. In other words, the path

$$x_1 \rightarrow x_{1-3} \rightarrow x_3 \rightarrow x_n,$$

is off-path pre-empted by the link

$$x_2 \not\rightarrow x_n$$

that is, in Figure 2.11, the path

$$x_1 \rightarrow x_2 \not\rightarrow x_n$$

is the preferred extension.

We may define *off-path pre-emption* as follows:

Definition 6 *Off-path pre-emption: In an inheritance network, if no redundant links exist or if the redundant link is interrupted by another node, a path that explicitly gives positive or negative information overrides nonexplicit paths (inherited properties).*

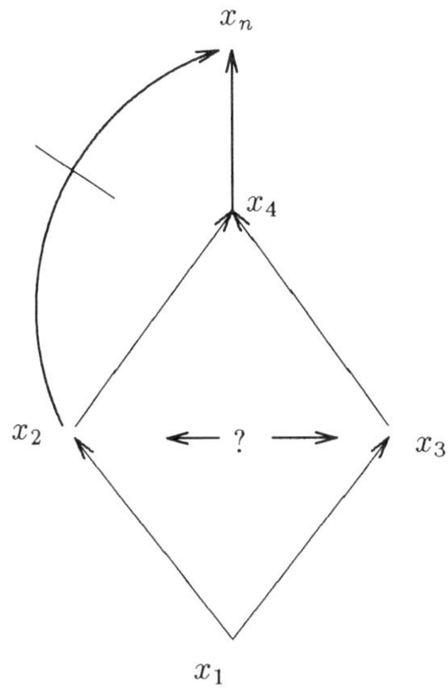

Figure 2.12: Γ3.

For example, consider Γ3 in Figure 2.12. The path

$$x_1 \rightarrow x_3 \rightarrow x_4 \rightarrow x_n$$

will be pre-empted by the path

$$x_1 \rightarrow x_2 \not\rightarrow x_n$$

if the latter contains more explicit information than the earlier, that is,

$$x_2 \not\rightarrow x_n.$$

2.5.1 Directions of Path-Based Reasoning

Given the different path-based mechanisms, Touretzky, Horty, and Thomason (Touretzky et al., 1987) highlighted the directions and the extensions that inheritance networks may follow. Whereas Touretzky, (Touretzky,

1986), using TMOIS, adopted the downward reasoning approach, others, such as (Horty et al., 1987) chose the upward reasoning approach.

Consider Figure 2.13. A skeptical downward reasoner could not draw conclusions regarding Clyde's property. This is because Clyde's related class (Royal.african.elephant) is ambiguous regarding the Grey.thing property in the first place.

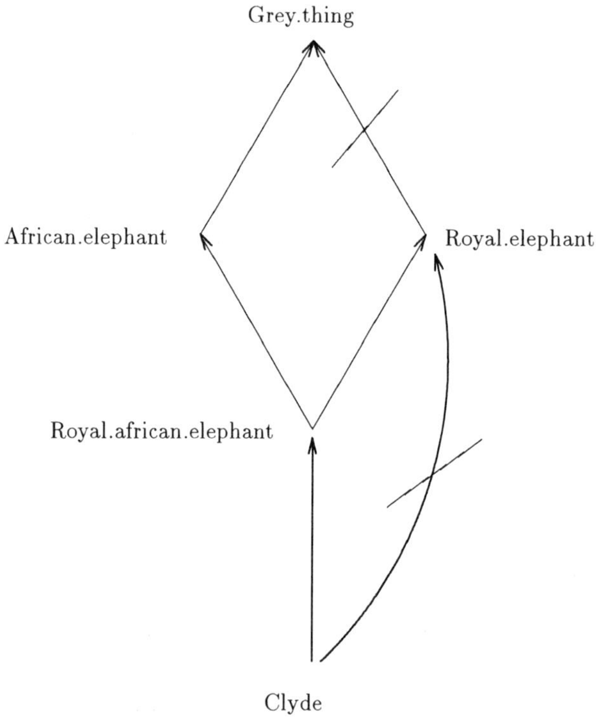

Figure 2.13: Downward versus upward reasoning.

On the other hand, in a skeptical upward reasoner, the link

$$Clyde \not\rightarrow Royal.elephant$$

rules out the inference path

$$Clyde \rightarrow Royal.african.elephant \rightarrow Royal.elephant \not\rightarrow Grey.thing$$

leaving the path

$$Clyde \rightarrow Royal.african.elephant \rightarrow African.elephant \rightarrow Grey.thing$$

to be the proposed extension. In such a case, Clyde is-a Grey.thing, even though its related class, Royal.african.elephant, remains ambiguous.

Alternatively, a credulous reasoner generates as many extensions as possible, so a credulous downward reasoner, of the example here, generates two extensions. One extension contains (ignoring the link $Clyde \not\rightarrow Royal.elephant$ for a moment):

$$Royal.african.elephant \not\rightarrow Grey.thing \qquad and$$

$$Clyde \rightarrow Royal.african.elephant \not\rightarrow Grey.thing$$

whereas the other contains

$$Royal.african.elephant \rightarrow Grey.thing \qquad and$$

$$Clyde \rightarrow Royal.african.elephant \rightarrow Grey.thing.$$

But because the extension

$$Clyde \rightarrow Royal.african.elephant \rightarrow Royal.elephant \not\rightarrow Grey.thing$$

is ruled out by the link

$$Clyde \not\rightarrow Royal.elephant$$

this network supports only the extension

$$Clyde \rightarrow Grey.thing.$$

Similarly, an upward credulous reasoner generates two extensions but because Clyde's property is independent of the Royal.african.elephant's property, as well as

$$Clyde \not\rightarrow Royal.elephant$$

Clyde will have the property Grey.thing in both extensions (Touretzky et al., 1987).

2.6 Inheritance Formula As A Representation Language

As discussed earlier, representing a great amount of information about real domains is essential to AI. Generally, real domains are imprecise and contain incomplete knowledge. Using inaccurate representation tools leads to difficulties concerning maintainability and expressiveness, for instance. For these reasons, a reasoning system must be able to manage and handle such domains through an appropriate knowledge representation formula for commonsense reasoning.

The formalizations of knowledge representation play the key role in specifying the best reasoning formula. Best formula here means deducing and inferring the right extension from existing information. However, two knowledge representation formalisms are considered to be equivalent if they share the same ability in acquiring the same amount of knowledge regardless of the different rules or the different mechanisms that have been used. For a general view of commonsense reasoning see, for example, (Davis, 1990).

As far as inheritance is concerned, the knowledge representation literature provides us with a variety of representation formulae to achieve this goal. However, although it is important to distinguish between the types of knowledge formalizations in representing facts and rules, such as path-based, logic-based, or lattice-based approaches, it is more important to discuss the goal behind the representation mechanisms in the first place and the way these formulae can achieve their goals.

To illustrate, earlier in the literature, logicians, such as Hayes, (Hayes, 1977), defined logic as a collection of ideas on how to express certain kinds of knowledge about certain kinds of world. Hayes (Hayes, 1979) tried to use semantic networks as a logical representation language. The translation between logic and semantic networks involves mapping constants to tokens (objects), unary predicates to types (kinds), axioms to links and a proof to path. Hayes did not consider the importance of various relations in different inheritance structures. In other words, all the inferences in his model are of the same importance.

On the other hand, Schubert (Schubert, 1976), for example, criticized the expressive weakness of semantic networks compared with predicate calculus, especially when handling quantification and higher order statements. Schubert distinguished between the expressive power of a language and its ability to deduce a new inference. He developed a new network representation that permits the use of n-ary predicates, logical connectives, unrestricted quantification in addition to the modal operators (such as belief), and counterfactuals implication. The representation accommodates propositions of the type encoded by Quillian (Quillian, 1968) in his networks.

Generally, classical first-order logic provides procedures, using usual transitive deductive reasoning, where no exception is allowed. Because exceptions are permitted in reality, both these approaches, among others, fail to supply the suitable inheritance representation formulae.

To overcome the shortcomings of classical first-order logic, different inheritance formalizations have been presented and different semantics, namely *proof-theoretic semantics* and *model-theoretic semantics*, have been proposed. Proof-theoretic semantics consists of providing algorithms and procedures for identifying acceptable paths supported by a network, whereas model-theoretic semantics consists of translating inheritance networks into some standard logical formulae and then generating model-theoretic interpretations of the formulae, (see Krishnaprasad & Kifer)(Krishnaprasad, 1989).

For instance, many logical formalizations that are modified forms of first-order predicate calculus (FOPC) have been presented, such as default logic (Reiter, 1980), circumscription (McCarthy, 1980), autoepistemic logic (Moore, 1984; Moore, 1985) and other formalisms. Unfortunately, inheritance problems involving pre-emption and contradiction do not get much attention. For instance, the conclusion of the statement, $Clyde \rightarrow elephant \rightarrow grey.thing$, is considered, logically, to be true, namely Clyde is a grey.thing by the logical rule of hypothetical syllogism (and not by default logic in particular). If Clyde happens to be a royal elephant, and royal elephants, typically, are nongrey.things, Clyde should be a nongrey.thing.

But the property grey.thing is no longer derivable from the superclass, elephant, because new information is added that retracts old information.

The problem with classical logic is that once we reach a conclusion we cannot then retract this conclusion even if it is later shown that it is false. Under such circumstances, we say that we have *nonmonotonicity*.

Classical logic (FOPC) cannot cope with nonmonotonicity because, as mentioned before, it uses rules of universal validity that can be applied to facts to deduce unretractable conclusions.

In order to overcome the shortcoming of classical logic, *Default Logic* (Reiter, 1980), among other candidates, presented some rule-based extension to the FOPC. Default logic's basic idea is that nonmonotonic reasoning derives default conclusions because not all the information is available and retracts these default conclusions when there is more information. However, there are major problems in its multiple inheritance representation[7]. What multiple inheritance structures require is a specific algorithm that can deal with Clyde in a more accurate and suitable way in order to get the right extension or the right theory.

On the other hand, there are several path-based approaches in the literature. Semantic networks, graphical representation, or tangled hierarchies are different names referring to the same subject, namely, knowledge representation using nodes and arcs. Quillian (Quillian, 1968) proposed an associational network model of semantic memory to capture the objective part of the meaning of words in formal representation. Quillian used nodes to represent word concepts and associative links to capture the semantic relation between them. His approach approximates the way people use their knowledge in representing events, abstractions, relations, hierarchies, and other different things.

Semantic networks have had a great influence in representing techniques. Semantic networks provide a natural way of representing inheritance relationships. For example, using a semantic network, inheritance relations can be created, supervised, modified, and updated more easily than in other representation languages. In addition, other representation languages, such as those based on logical representations, lack the ability to handle important features of inheritance, such as multiple inheritance structures and inheritance with exceptions, in proper and suitable ways.

The path-based approaches have some general advantages over logic-based approaches. According to the nature of inheritance, logical approaches, in general, miss specific semantic explanations. The second point is that all logical formalizations follow the credulous approach, in which all the extensions have the same potential to appear. That is,

[7] I will present this kind of nonmonotonic reasoning, among others, in detail in the next chapter.

there is no basis in preferring unique extensions, whereas proof-theoretic approaches offer a more accurate method in preferring one extension offered by a structure. Finally, logical based approaches, because of their 'flat' textual structure and lack of modularity, cannot deal with the topology of different inheritance structures that are offered by different hierarchies. Translations of inheritance structures into logic do not preserve the visual structuring offered by the path-based approach.

In spite of the advantages of path-based over logic-based approaches, earlier inheritance representation systems, such as NETL (Fahlman, 1979), TMOIS (Touretzky, 1986), and KRL (Bobrow and Winograd, 1985), have no specific procedure for dealing with inheritance structures with exceptions. That is, when exceptions do exist within an inheritance hierarchy, they try to deal with pre-emption and contradiction by either preferring one class to others in an ad hoc way, or giving no conclusion in order to avoid inconsistency.

For example, NETL (Fahlman, 1979) handles the redundant information problem by sending markers from the related classes to a specific property. NETL chooses the class that reaches its property goal earlier, that is, the shortest length among the computing marker propagation paths. Obviously, search algorithms that determine inherited properties based on the *length* of competing inference paths are inadequate for this task. We must use some ordering procedure that allows one class to override others if they are in a proper sub-superset relation, or else prevents redundant information from being entered.

TMOIS (Touretzky, 1986) introduces a kind of topological relation called inferential distance ordering (IDO); (see Chapter 4).

What we need is an interpretation procedure in accord with human intuition concerning the meanings of inheritance structures, such as how to make the right decision between conflict classes.

What I want to stress here is that to represent real-world knowledge we need to deal with our intuition concerning the behavior and interpretation of inheritance structures rather than just typical default properties, that is, we need to believe, if we have been told that Clyde is black-elephant, that Clyde is a black-thing, even if Clyde cannot be related to that class by default.

In this volume, new concepts are presented that lead to an investigation of four areas:

1. Exceptional inheritance reasoning, (EIR), for pre-emption problems.
2. Default correlation (DC) for contradiction problems.
3. Defining the relation between inheritance reasoning and causal reasoning.
4. Defining the relation between inheritance reasoning and analogical reasoning using causal reasoning.

As explained in Chapter 5, Exceptional Inheritance Reasoning provides a novel way of overcoming the shortcoming of IDO by treating problems as they arise in structures involving redundant information. EIR achieves this by identifying an implicit, missing relation and adding it explicitly to the network. EIR reduces the problematic role of redundant statements and makes explicit implicit relations between nodes in inheritance networks.

Chapter 3

Current Approaches to Nonmonotonic Reasoning

3.1 Introduction

In order to create plausible intelligent systems that can reason about the real world, a wide range of reasoning styles have been proposed. The technical problem with these proposals lies in their inability to describe human commonsense reasoning precisely or intuitively. The real world is dynamic, with new states of affairs arising out of old. Such new states replace old states of affairs, and new states must be represented in such a way that old states are cancelled out. Monotonic reasoning styles assume an incremental world, where new states and situations are continually added to the existing states and situations, but the real world does not operate in this manner. Any reasoning style that fails to take into account the simple fact that a new state may cancel out (i.e., contradict) a previous state will be severely limited in its ability to capture real-world information or represent the dynamic nature of reality. Nevertheless, most work on reasoning styles in AI has focused on monotonic systems. This would be surprising if it is forgotten that one of the dominant paradigms in AI is the use of logic-based reasoning styles, where for instance the derivation of a new fact is like deriving a theorem. In standard logic reasoning styles, there is no such thing as generating a possible fact and then checking its truth. To derive a possible fact is to derive a theorem, given the truth of the premises. Any such theorems can be thrown back into the melting pot of existing facts, and further theorems can be derived with the added theorems. There must be no contradiction between later theorems and earlier theorems and premises in a monotonic system. Any reasoning style that attempts to evaluate new facts in terms of their effect on existing facts, so that statements that are no longer true can be removed to leave a consistent system state, is a nonmonotonic system.

It is a common view that nonmonotonicity, being a property that can be represented logically (a system is monotonic if, from a set of premises with a conclusion, that same conclusion can be derived from every superset of the original set of premises), must be tackled logically. Most literature on nonmonotonic systems assume that an extended logic approach (i.e., first-order predicate calculus with extensions, such as modality, defaults, probabilities, fuzziness, etc.) is sufficient. However, there is an alternative approach. This chapter is an introduction to the two main approaches to nonmonotonic reasoning.

3.2 Path-Based Review Literature

Let us start with the path-based approach, which is the main alternative to the logic-based approach.

We may consider NETL, a system developed by Fahlman (Fahlman, 1979) as the starting point for our examination of multiple inheritance with exceptions. NETL is a parallel machine that represents real-world knowledge. Fahlman's goal was to construct a system that makes possible the efficient deduction of implicit information from explicitly asserted facts in a knowledge base, irrespective of size. Fahlman believed that this goal could be achieved by using a parallel network. In such a network, node-units will represent concepts (entities) and correspond to processors in the knowledge base, whereas link-units will represent relationships between the various nodes and correspond to hard-wired connections between processors. These units are hardware devices that propagate marker bits from node to node and in parallel. Propagation was under the control of a network controller. NETL uses links to denote facts and defaults between concepts. Crossbar links are used to denote negative relations between such concepts. To infer implicit relations between two processors, NETL passes markers along the links that relate these processors in the network. In a conflict situation and in order to choose one path among other paths between two nodes (multiple inheritance with exceptions), NETL chooses the marker that arrives at the end node first; that is, it chooses the shortest path between two nodes. This approach is called *shortest path inference*.

Touretzky (Touretzky, 1984) introduced inferential distance ordering – an implicit ordering rule that is able to handle exceptions in multiple inheritance structures. In his approach, Touretzky (Touretzky, 1986) highlighted the differences between semantic networks and the default logic of Reiter (Reiter, 1980) when representing inheritance structures. Touretzky criticized default logic by emphasizing that, as information is added to a knowledge base, existing default rules need to be replaced by new rules that take the new exceptions into account. Second, as the knowledge base grows, the complexity of each individual default rule will increase as the number of exceptions grows. Finally, in any given inheritance network, the translation of a link cannot be determined independently of the other links. Default logic provides representations only after the inheritance network has been interpreted, in other words, and the interpretation cannot obviously be a default logic one, because this is the representation we are trying to construct in the first place. Default logic by itself cannot interpret semantic network inheritance structures. Also, Touretzky claimed that default logic does not preserve the explicit (visually interpretable) structure of the inheritance network and therefore cannot represent implicit default ordering. Topological Inheritance Architecture, (TINA); (Touretzky, 1984), an inheritance reasoner based on inferential distance ordering, constructs the extensions of unambiguous inheritance networks by incrementally generating inheritance paths and weeding out those that violate the inferential distance ordering. Touretzky applied inferential distance to default logic by using ordering on proof sequences over the set of possible extensions. Touretzky (Touretzky, 1986) used The Mathematics Of Inheritance System (TMOIS); to describe the implicit aspects of inheritance structures, and other definitions for inheritance have since been proposed by (Sandewall, 1986), (Brewka, 1987), (Horty et al., 1987), (Etherington, 1987b), (Neufeld, 1991), and (Al-Asady, 1993a) that present different path-based approaches.

Horty et al.(Horty et al., 1987) described a 'restricted' skeptical (defeasible) approach to path-based inheritance. The importance of their approach lies in using the bottom-up approach for path concatenation.

Whereas previous approaches use a top-down approach (Fahlman, 1979), (Touretzky, 1986), that is, paths are formed from the root or most general class down to more specific classes and finally individuals, the bottom-up approach forms paths (*compound paths*; (Horty et al., 1987)) from individuals and more specific classes up to the root or most general class.

The difference between the two is that, in the top-down approach, 'coupling' is preserved; that is, an instance or subclass inherits everything that its class inherits (plus adds its own specialized information). With uncoupled (bottom-up) inheritance path concatenation, an instance or subclass possibly does not inherit everything its class inherits (see Chapter 2 for the definitions of the different types of concatenation).

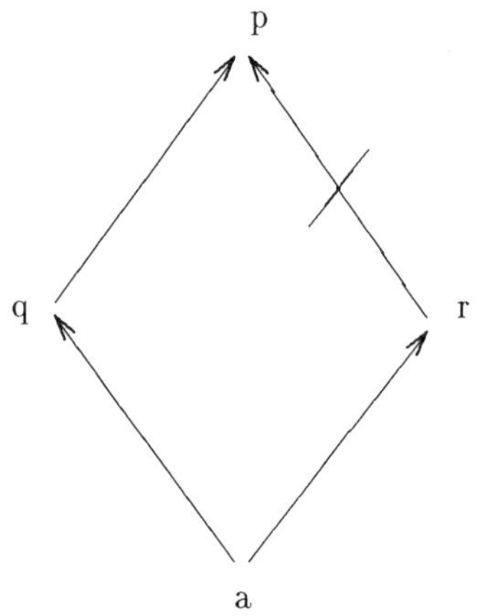

Figure 3.1: Γ3, compound conflicting paths (Horty *et al*, 1987).

To illustrate, Horty, Thomason and Touretzky (Horty et al., 1987) distinguished between three types of chaining, using Figure 3.1. *Unrestricted forward chaining:* constructs the path $a \rightarrow q \rightarrow p$ *and* the path $a \rightarrow r \not\rightarrow p$ at the same time (Touretzky, 1986). The second type is the standard *credulous approach* which constructs *either* the path $a \rightarrow q \rightarrow p$ *or* the path $a \rightarrow r \not\rightarrow p$ at the same time. The third type, which is supported by Horty et al. (Horty et al., 1987) is the *restricted skepticism approach*. Restricted skepticism supports *neither* the path $a \rightarrow q \rightarrow p$ *nor* the path $a \rightarrow r \not\rightarrow p$ in the net according to the idea that arguments tend to neutralize each other. In other words, each extension in a net is supported by a path and has an argument to the contrary. Using the skeptical approach, and in order for a net to support one single path, an intermediate node on one of the conflict paths with different polarity directed to the end point is needed.

To clarify this consider Figure 3.2 and the following paths:

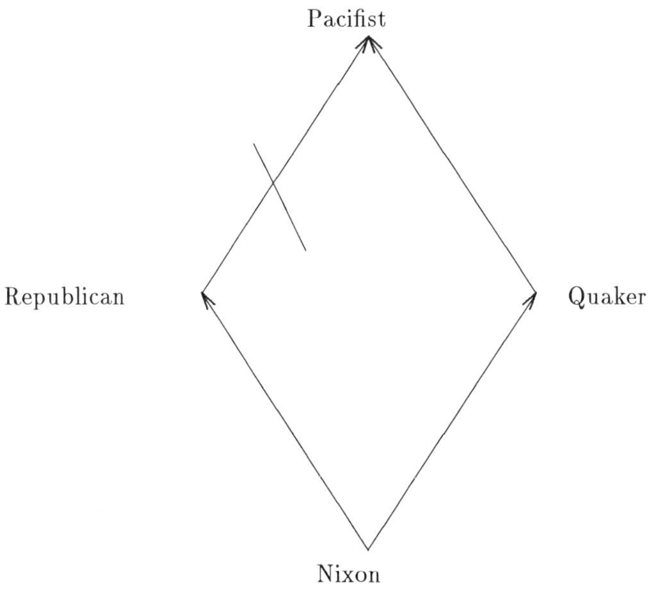

Figure 3.2: Nixon Diamond problem.

$$Nixon \rightarrow Republican \text{ and } Nixon \rightarrow Quaker,$$

$$Republican \not\rightarrow Pacifist \ , \ Quaker \rightarrow Pacifist.$$

The path *Nixon* $\rightarrow$ *Republican* $\not\rightarrow$ *Pacifist* is not inheritable because there is a conflict path, namely *Nixon* $\rightarrow$ *Quaker* $\rightarrow$ *Pacifist*, provided that no path from Nixon to Quaker with an intermediate node X, such that $X \not\rightarrow$ *Pacifist* does exist.

However, we may consider this skeptical approach as an intersection of all credulous extensions. In fact, this is not the case according to Horty et al. (1987):

> It is tempting, therefore, to suppose that the set of paths permitted by a given net under restricted skeptical analysis might simply be the *intersection* of the various extensions associated with that net according to the credulous analysis provided by (Touretzky, 1986). However, nets like Γ14 (Figure 3.3) –which have the topology of *nested* Nixon Diamonds– show that this is not the case always. (p. 330)

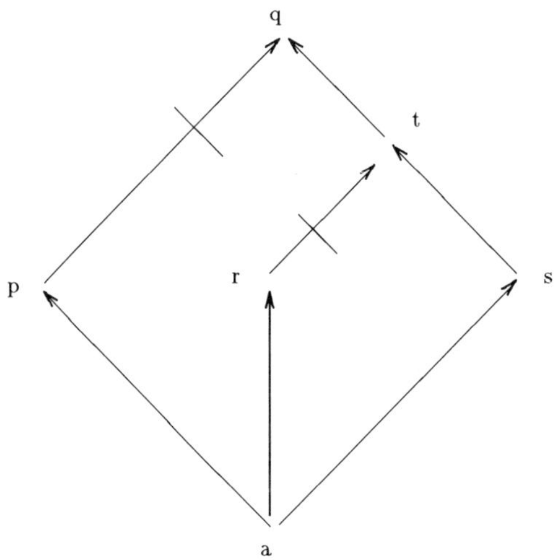

Figure 3.3: Γ14, intersections of credulous extensions (Horty *et al*, 1987).

In other words, Horty et al., using Figure 3.3, explained that when using the credulous approach, two extensions will be supported: the extension $a \to s \to t \to q$ and the extension $a \to p \not\to q$. But when using the skeptical approach, because the initial segment $a \to s \to t$ of the path $a \to s \to t \to q$ is neutralized by the path $a \to r \not\to t$, the net $\Gamma 14$ will support the path $a \to p \not\to q$ because the potentially conflicting path $a \to s \to t \to q$ poses no problem. $\Gamma 14$ does not support the path $a \to p \not\to q$ in all its associated extensions (according to the credulous approach) because some extensions contain instead the path $a \to s \to t \to q$. In the presence of multiple ambiguities, skeptical inference may permit certain paths that are not permitted in all credulous extensions (Touretzky et al., 1987).

To solve ambiguity from a skeptical point of view two inheritance approaches have been adopted: *ambiguity blocking inheritance*, which disallows further ambiguous lines of reasoning as soon as an ambiguous node has been reached (this is the approach proposed by Horty et al. (Horty et al., 1987)), and *ambiguity propagation inheritance*, which allows ambiguous lines of reasoning to interfere with ambiguous situations (this is the approach proposed by Stein (Stein, 1989)). The differences

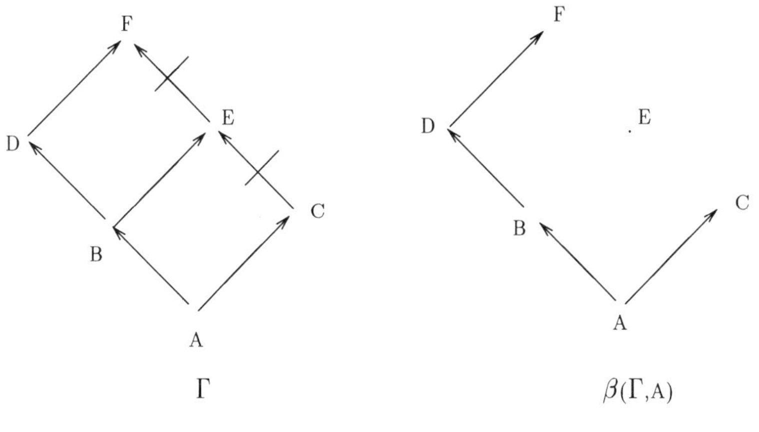

Figure 3.4: Applying ambiguity blocking inheritance to Γ yields $\beta(\Gamma,A)$ (Stein 1989).

between the two approaches may clarified more using the following ex-

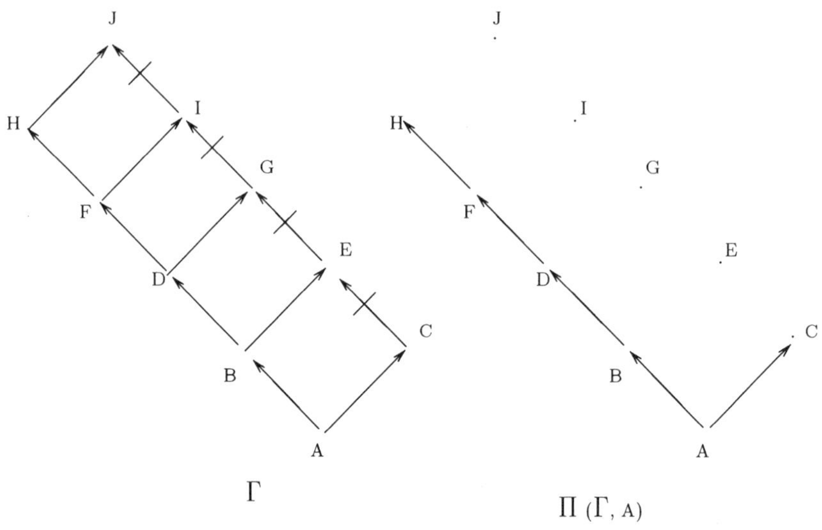

Figure 3.5: Applying ambiguity propagation inheritance to Γ yields Π(Γ,A) (Stein 1989).

amples and Figures 3.4 and 3.5 ((Stein, 1989)).

Given an inheritance network (in Figure 3.4), Γ, we need to decide whether A will inherit E or F because these nodes are ambiguous to A. *Ambiguity blocking inheritance* on Γ determines that E is ambiguous with respect to the node A, so it eliminates all relations *to and from* the node E leaving the node F unambiguous with respect to A as shown in the net $\beta(\Gamma, A)$.

Ambiguity blocking inheritance decides this view, but *ambiguity propagation inheritance* has a different view here. Given an inheritance network (in Figure 3.5), Γ, *ambiguity propagation inheritance* draws no conclusions about whether A is-a E or G or I or J because this type of reasoning allows ambiguous lines of reasoning to proceed unless there is no counterargument. In Figure 3.5, $\Pi(\Gamma, A)$ never supports ambiguous nodes.

Horty et al., (Horty et al., 1987) defined a skeptical approach in that

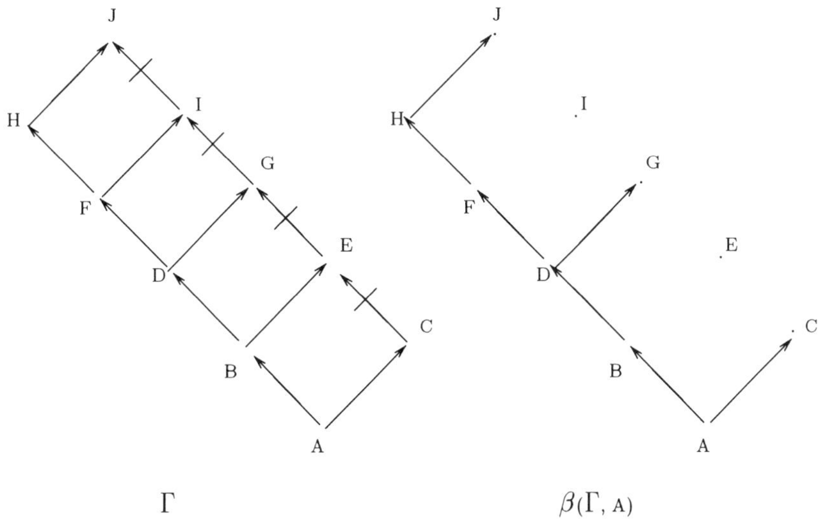

Figure 3.6: $\beta(\Gamma, A)$ (Stein 1989).

it did not propagate ambiguities in the presence of multiple ambiguities (there may be skeptical conclusions not contained in all credulous extensions of an inheritance network), but Stein (Stein, 1989) criticized this approach using Figure 3.6. According to the Horty et al. approach, although $\beta(\Gamma$, A) supports the extensions: $A \rightarrow C, A \rightarrow G, A \rightarrow J$, it does not support the extensions $A \rightarrow E, A \rightarrow I$. Also, in addition to supporting the extension $A \rightarrow J$, it supports the extensions $D \rightarrow J, H \rightarrow J$ as well, but it gives no idea whether $B \rightarrow J$ or $F \rightarrow J$.

A second criticism to Horty's approach lies in the skeptical definition. Skeptical definition is based on the intersection of all extensions and is therefore not sufficient for considering any path contained outside the intersection nodes. Stein criticized this definition as follows:

Consider *either* the path $A \rightarrow X_1 \rightarrow Y$, or the path $A \rightarrow X_2 \rightarrow Y$ is contained in all credulous extensions, but none of them is considered in all these extensions. It is certain that the inference $A \rightarrow Y$ is the right extension, even with no path supporting this conclusion.

In her approach, Stein (Stein, 1989) described the *ideally skeptical inheritance* approach as follows. Ideally skeptical inheritance adopts ambiguity propagation after computing the inferences that have some supporting argument in every credulous extension of the structure. Ambiguity propagation can be obtained by marking the ambiguous nodes in a net, so that these nodes will be outside the conclusion decision but they have an antithesis role to play in preventing other nodes from being unambiguous. Stein argued that the importance of reasoning in some inferences that are supported by every credulous extension should go through inferences (conclusions) rather than path-based reasoning arguments. This may be clarified by using Figure 3.7 and the following example ((Stein, 1989)):

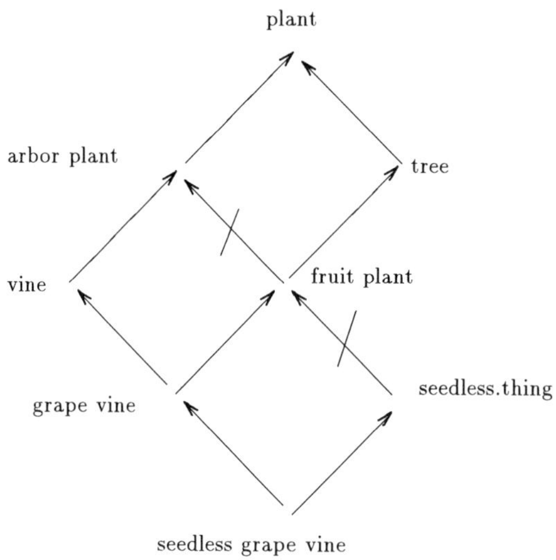

Figure 3.7: Γ3, (Stein89)

Although the conclusion *seedless grape vine* → *plant* is supported by every credulous extension in Γ3 (Figure 3.7), an ideally skeptical approach will support this conclusion as well. A seedless grape vine is a plant whether or not it is a fruit plant (because it is unambiguously an arbor plant).

> 'In any state of the world, no matter how we resolve the ambiguities of the taxonomy, a *seedless grape vine* is a *plant*' (p. 1156).

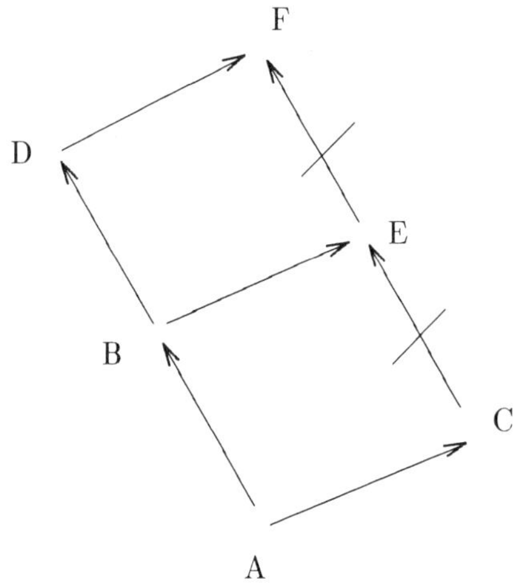

Figure 3.8: Γ4, Ideally skeptical inheritance.

In summary, whereas the skeptical approach (Horty et al., 1987), using Γ4 of Figure 3.8, supports the conclusion $A \rightarrow F$, an *Ideally skeptical inheritance* supports *neither* the conclusion $A \rightarrow F$ *nor* the conclusion $A \not\rightarrow F$ because Γ4 has credulous extensions supporting either conclusion. I.e., Stein's *Ideally skeptical inheritance* is more restricted than the Horty et al. skeptical approach.

In another approach, Thomason, Horty, and Touretzky (Thomason et al., 1987) presented a paper of a project devoted to interactions between logical theory and (monotonic) inheritance networks.

In this approach, Thomason et al. defined an inheritance relation, $\models$, as follows: If a statement A is supported by a net Γ, we say that $\Gamma \models A$. Γ permits a monotonic path that supports A. Also they define $\vdash$ as follows: $\Gamma \vdash A$ means that a set of formulas, Γ, has enough information to yield the formula A.

Thomason et al. started by criticizing the belief assumption that monotonic inheritance networks [1] are semantically equivalence to classical logic, where

$$a \rightarrow p \text{ is translated as } pa, \qquad a \not\rightarrow p \text{ as } \neg pa,$$
$$p \rightarrow q \text{ as } \forall x[px \rightarrow qx] \qquad \text{and} \qquad p \not\rightarrow q \text{ as } \forall x[px \not\rightarrow qx].$$

For example, consider the net $\Gamma = \{a \rightarrow p, a \not\rightarrow p, a \rightarrow q\}$. Using classical logic, Γ will support the sequent $a \rightarrow p, a \not\rightarrow p \vdash a \rightarrow q$, (from inconsistent premises we can infer anything), whereas a semantic network shows $\{a \rightarrow p, a \not\rightarrow p\} \not\models a \rightarrow q$.

Thomason et al. (1987) wrote:

> However, this is wrong, even in the case of monotonic inheritance. Examples such as *the above* seem to have been generally overlooked. In fact, it seems to be a kind of 'folk theorem' in artificial intelligence that the logic of semantic networks, and even frame systems, is just the classical first-order predicate calculus. (p. 281).

This led Thomason et al. to define the semantics of monotonic inheritance network in terms of four valued logic. Although Hayes' paper (Hayes, 1979) is usually regarded as one of the earliest papers to attempt to formalize Minsky's frames (without exceptions) (Minsky, 1975), Brewka's approach (Brewka, 1987) may be considered as the first attempt to show how the semantics of frames, including exceptions, could be handled using formal logic. Brewka's approach deals with frame systems (a strict multiple is-a hierarchy as well as defeasible links-slots-which represent attributes and point from classes or frames to typical attribute values). A simple frame language for multiple inheritance is used (because frame systems do not use graphical notation). A new three-place predicate called *Exceptional* is introduced using variable circumscription (McCarthy, 1980). The expressions of this language can be translated to first-order formulas according to the way that the predicate *Exceptional* can be circumscribed. In his system, Brewka used the

[1] Where all links represent true statements.

reification technique[2]. For example, instead of writing MAN(Peter) to express the fact that Peter is an instance of the frame MAN, we introduce a predicate IS and a constant Man and express this fact in first-order logic as IS(Peter,Man) (Brewka, 1987). The following example contains two frame definitions and one instance definition of how this works is taken from (Brewka, 1991).

(defframe CAR
 (slots (WHEELS 4)(SEATS 5)))
(defframe SPORTCAR
 (supers CAR)
 ((slots (SEATS 2)(CYLINDERS 6)))
(definstance SPEEDY of SPORTCAR with CYLINDERS = 8).

Now to translate this frame language into a set of first-order formulae, a three-place predicate EXCEPTIONAL of the form

EXCEPTIONAL(x,$Slot_1$,$Frame_1$) is used as follows:

IS(SPEEDY,SPORTCAR)
 & EXCEPTIONAL(SPEEDY,CYLINDERS,SPORTCAR)
 → HOLDS(CYLINDERS,SPEEDY,8)

which demonstrates no inconsistency in the given translation. It turns out that the meaning of a Brewka's frame system can be defined in a logical way rather than by resorting to Touretzky's path-based approach. In addition to this, Brewka discussed ambiguity in frame systems in which the intuition in this system is skeptical if no additional information is known.

Horty and Thomason, (Horty, 1988) emphasized that, within inheritance structures without the ability to express the strict and the defeasible information together, an inheritance system cannot represent definitional relations among concepts. In (Thomason et al., 1987) the authors presented an inheritance theory that deals mainly with strict links and claim that a strict network permits exactly the strict path it contains[3]. In (Horty et al., 1987) the authors presented an inheritance

[2]Brewka explained that second-order logic would be a good formalism to express the specialization principle if frames are interpreted as predicates in that one frame will specialize another, but second-order logic introduces many difficulties. For that reason he used constants instead of predicate symbols, which allow him to reason about properties of frames in first-order logic (Brewka, 1987).

[3]Using strict inheritance network, $\Leftarrow$ *and* $\Rightarrow$ represent positive relations and $\not\Longleftrightarrow$ represents negative relation (Horty, 1988).

theory that deals with defeasible information. The authors explained that by using the skeptical defeasible approach, the conflict arguments, within an inheritance network, tend to neutralize each other. In (Horty, 1988) the authors explored a new mixed theory in which strict and defeasible (positive and negative) relations can be combined in a single inheritance network.

Dealing with networks containing strict links alone is not a complex matter. However, problems arise when a defeasible path is included in the structure. In order to combine these two theories into an account of inheritance for mixed networks, the authors carry over the analysis of strict inheritance. Because a strict network permits exactly the paths it contains, it creates no problem. But the problem is with defeasible networks because they do not permit all the paths they contain, but direct links (ordinary path), or any other path with which it conflicts is itself pre-empted. For those reasons, a modification to the concept of the kind of paths that represent conflicts is done as well as a new definition to the pre-emption relation is created.

Generalized path, a new terminology is created. A generalized path is a sequence of links joined like an ordinary path, except that it can contain negative segments anywhere. For example, $p \not\rightarrow q \Leftarrow r \not\rightarrow s \Leftarrow t$ is a generalized path (Horty, 1988).

On the preemption side, the authors assigned a measure of complexity to each path σ in such a way that it can be decided whether a net supports σ or not. A path σ is divided into subpaths $v(\sigma)$ and $\delta(\sigma)$, where $v(\sigma)$ is the maximal strict end segment of σ, and $\delta(\sigma)$ is the result of truncating $v(\sigma)$ from σ. For example, if σ is $x \Rightarrow y \rightarrow p \not\Longleftrightarrow r \Leftarrow s$, then $v(\sigma)$ is $p \not\Longleftrightarrow r \Leftarrow s$ and $\delta(\sigma)$ is $x \Rightarrow y \rightarrow p$) (Horty, 1988).

Using this, the authors specified the conditions under which a net could support σ in a mixed inheritance theory.

Figure 3.9 is an example of a mixed strict and defeasible inheritance network. In Γ5, neither $p \rightarrow q \rightarrow r$ nor $p \rightarrow u \rightarrow v \not\rightarrow t$ will be permitted because each is neutralized by its conflict with the other.

Selman and Levesque (Selman, 1989) indirectly summarized the basic differences between the logic-based approach and the path-based approach. Selman and Levesque's argument was that, in order to distinguish between the different proposals arising from path-based systems, computational tractability can be used as a basis for comparison. The

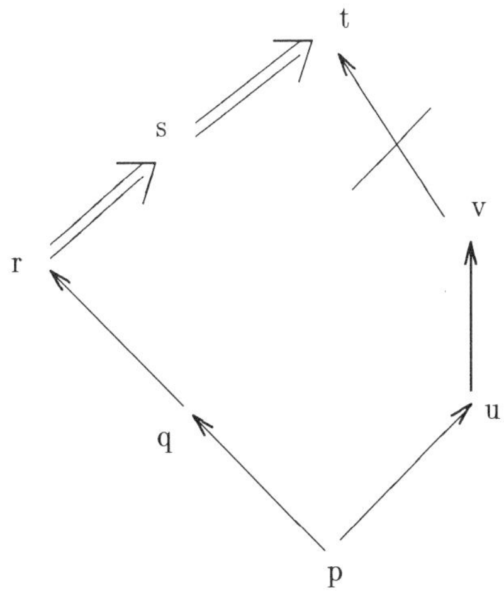

Figure 3.9: Γ5, mixed inheritance structure, (Horty et al. 1988).

main finding of their paper was that pre-emptive-based reasoning, including Touretzky's TMOIS (Touretzky, 1986) is NP-hard (i.e., the algorithms that exist are nonpolynomial and, by implication, exponential depending on the number of nodes in the graph). This in turn implies that the 'conditioning' of a network is NP-hard before parallel inferencing can occur:

> 'This suggests that a Touretzky inheritance procedure (TMOIS) cannot run unsupervised, unless the network can be restricted in form or in size' ((Selman, 1990), p. 94).

One important aspect of this is the overview it provides of path-based approaches: Basic graph definitions are introduced before contradiction, on-path intermediate, off-path intermediate, pre-emption, upward and downward concatenation, inheritable path, credulous and skeptical grounded extension, and the inheritance problem from a path-based point of view are all addressed. One important result of Selman's (1990) analysis is that it is only upward reasoning (concatenating nodes in a path bottom-up) that is tractable, but upward reasoning has the undesirable computational complexity of uncoupled inheritance: A descendant of a parent need not necessarily possess the same inherited information as its parent inherited from its own parents. All inheritable information in an uncoupled system is therefore default unless, for instance, some way is found of distinguishing default from nondefault information and then providing different inheritance mechanisms for dealing with both types of inheritable information.

Ginsberg (Ginsberg, 1991) criticized existing formalizations of inheritance reasoning in that these approaches are considered to deal with just a specific (local) domain area. The work by Horty (Horty et al., 1987), and Touretzky (Touretzky et al., 1987) are examples is the claim. There is also no way for these systems to predicate the declarative translation of a new link after adding it to an inheritance structure without examining the remainder of the diagram. This drawback is found in the works of Etherington and Reiter (Etherington and Reiter, 1983), Fahlman (Fahlman, 1979), Krishnaprasad (Krishnaprasad and Kifer, 1989), and Touretzky (Touretzky, 1984), for example. Ginsberg presented a new local formalization of inheritance reasoning using causal arguments of the type 'if a normality assumption about a superclass causes an abnormality with regard to a subclass, then the original normality must have been in error'. Ginsberg applied his proposed formalization to a variety of inheritance structures and examined it by showing its similarity and differences from other approaches.

Geffner and Verma (Geffner, 1989) presented a new path-based inheritance approach. Although its structure is influenced by the approach presented by Horty et al. (Horty et al., 1987), this approach is different in that it embodies a more appropriate notion of specificity and defeat. The authors defined new structural notations used to identify two inference rules: The first defines the link the reader can understand as a replacement to an axiom, and the second defines the chain that is the derivation or the extension of a network. Although it turns out that the complexity of this system is NP-hard, the proposed algorithm that computes some of the inferences provides a sound but incomplete approximation and has the ability to handle some examples correctly.

3.3 Logic-Based Approach

In this section I highlight the different logic-based approach to redundancy and ambiguity. I start with monotonic views, then move to nonmonotonic approaches.

3.3.1 Monotonic Logic View

Although Woods (Woods, 1975) described the importance of semantic network representations in human belief and understanding, he criticizes semantic nets by questioning the role of a link in various representation schemes. In other words, what is the meaning of the link, and are semantic nets suitable for representing real-world knowledge? Woods argued that it is necessary to represent intentional concepts, in the sense in which, for example, morning star and evening star are intentionally distinct concepts, although referring to the same extension (the planet Venus), (Woods, 1975). However, Woods argued that such intentions cannot be expressed in first-order logic.

Hayes (Hayes, 1977; Hayes, 1979) defended logic as an adequate representation language. Hayes, like Woods, criticized semantic nets as a kind of formalism without a well-defined semantic interpretation, which makes it weaker than logic, expressively and deductively. Hayes argued that inheritance hierarchies can be represented and expressed by including references to belief states. However, he did not show the characteristics and types of structure that could be represented.

However, the shortcomings of classical logic when dealing with exceptions and changes over time were addressed by the appearance of different modified versions of classical logic, designed specifically for nonmonotonic reasoning, such as default logic, nonmonotonic logic, circumscription and other extended versions of these logical calculi.

3.3.2 Nonmonotonic logic view

'Nothing is certain, but death and taxes'

(Benjamin Franklin 1789)

A central concern in AI is how to define a formula with an ability to revise previous conclusions when more information to the contrary arrives without retraction of previous, true information. To do so, many logical approaches are proposed. These approaches provide traditional model-theoretic semantics (formal semantic models) for a structure by encoding the structure in some existing logic. Many other proposals have appeared in the literature explaining either a modified version or an extended version of these formalizations. Because the literature concerning this subject is vast, I will adoe main logical approaches concerning inheritance structures.

3.3.3 Modal Nonmonotonic Logic

One of the earliest consistency-based approaches is McDermott and Doyle's nonmonotonic logic (McDermott and Doyle, 1980). McDermott and Doyle defined nonmonotonic logic I (NMLI) and McDermott (McDermott, 1982) defined nonmonotonic logic II (NMLII) to overcome the problems with NMLI.

In general, NML is a logic that uses a modal operator M in conjunction with a first-order logic to represent defaults. For example, 'MA' stands for 'A is consistent with everything believed'. Default statements are formally treated as formulas in a theory and not as rules of inference, thus formulas in a theory need to be consistent with each other. For example, NML represents the default 'birds fly' by the statement:

$$(\forall x)((Bird(x) \wedge MFly(x)) \supset Fly(x))$$

which can be read as 'for every x, if x is a bird, and it is consistent that x flies, then x flies'. Here is another example ((McDermott and Doyle, 1980)). If the following assertions are introduced:

(1) noon $\wedge$ M[sun-shining] $\supset$ sun-shining,

(2) noon,

(3) eclipse $\supset$ $\neg$ sun-shining.

the following fixed point[4] could be proven:

(4) sun-shining.

If we add the following axiom

(5) eclipse, then (4) can not be proven; that is (4) is not a theorem of the extended theory.

It is possible to assert that (4) is consistent with a theory and that (4) is false. M(sun-shining) is not inconsistent with ¬(sun-shining); so NML has been shown to be inconsistent.

3.3.4 Autoepistemic Logic

Moore (Moore, 1984; Moore, 1985) adopted autoepistemic logic (AEL). AEL is a more extended and more successful proposal than McDermott's NML. AEL can be defined as a logic of an agent's own belief (rather than consistency), such as 'if Losy is a cat and if you don't believe that Losy is not a mammal, then Losy is a mammal' (Moore, 1984)[5].

Belief knowledge is monotonic, such as 'if I do not believe F, then T is true.' To achieve nonmonotonicity (in order to allow an agent to retract his own beliefs), we need to add F to an agent's premises, then T will no longer be derivable; that is nonmonotonicity achieved by sanctioning the agent's own beliefs.

Instead of formalizing the concept of consistency (McDermott and Doyle, 1980), Moore presented a formal account of an ideal agent reasoning about his own beliefs in the sense that he knows that he knows X, whenever he knows X, and he knows that he does not know X whenever he does not know X.

Using AEL, the modal operator M has been given a different interpretation in which it expresses a relationship between what is believed and what is true, that is, the modal operator L is read as 'it is believed

---

[4]Fixed point corresponds to belief sets that can be obtained by applying the standard inference rules of classical logic and throwing in as many formulae of the form M_p as possible.

[5]Epistemic logic is a special kind of intentional logic that represents intentional verbs (e.g., know, believe) among others (e.g., think, hope, wish, fear, and imagine). An intentional verb is one that express some mental attitude toward a proposition (Sowa, 1992).

that' whereas the operator M is read as 'it is not believed that not'[6]. For instance we might represent the earlier example as

$$Cat(Losy) \wedge \neg L \neg Mammal(Losy) \supset Mammal(Losy).$$

However, unlike Moore's AEL and other versions of epistemic logic, Konolige introduced a modification of AEL called Hierarchic Autoepistemic Logic (HAEL). Konolige et al., (Konolige, 1988b; Konolige, 1988a; Konolige, 1989), identified two of the main problems for nonmonotonic logics: the lack of computability (tractability) and ambiguity. HAEL's rules link what is believed (known)[7] with what is true. It allows the representation of priorities between defaults. The basic idea is that each assertion or rule is given a subtheory label. Then the operator L in each subtheory refers to theories that are lower in the hierarchy of subtheories. This means that an ordering on subtheories is required. For instance, $L_{i}p$ refers to the default p that is believed in subtheory T_i of T.

Brewka (Brewka, 1991) introduced *preferred subtheories*, in which he identified possible objections to HAEL. First, it forces the expression of preferences rather than leaving the expression of preferences as a possibility. This is a problem where there may be no information about what the priorities should be between different defaults and thus theories. With regard to the implementation of HAEL, Brewka criticized HAEL in that every rule that has defaults may need its own theory. Although this may be computationally attractive in terms of tractability, the cognitive implication that there may be as many different levels of belief as there are rules needs addressing.

3.3.5 Default logic

Reiter (Reiter, 1980)introduced default logic. Reiter's default logic interprets the statement 'A's are typically B's' as 'if x is an A and it is consistent to assume that x is a B, then x is a B'. Default statements are formally treated as *meta-level rules of inference.*

[6] Using modal logic, M_p stands for 'p is possible' or 'p is consistent', L_p stands for 'p is necessary' or 'p is provable', L_p can be expressed as $\neg M \neg p$ (Brewka, 1991).

[7] In many applications of logic to AI the philosophically important distinction between knowledge and belief is often blurred.

A default theory (Reiter, 1980) consists of a set, W, of ordinary first order formulae together with a set, D, of rules of inference called *Defaults* which have the form:-

$$D \qquad \frac{\alpha(x) : \beta(x)}{\gamma(x)}$$

where α, β, γ are first order formule whose free variable is x, α is prerequisite, β is justification, and γ is consequent. For any individual x, if $\alpha(x)$ is inferrable and if $\beta(x)$ is consistent, then $\gamma(x)$ can be inferred.

For instance we may interpret the following: 'if Clyde is an elephant is given, and 'Clyde being grey' is consistent with what else is currently known, then we can conclude Clyde is grey' as follows:

$$\frac{elephant(Clyde) : grey(Clyde)}{grey(Clyde)}$$

To formalize inheritance hierarchies with exceptions, Reiter and Etherington focus on the semantics of default logic (Etherington, 1987a; Etherington, 1987c).

A default is said to be *normal* if the consequent is the entire justification, that is $\beta(x)$ and $\gamma(x)$ are identical.

However, because nonmonotonicity is caused by the presence of exceptions, normal default logic is not adequate for representing multiple inheritance and exceptions. For example, if an individual, Henry the ostrich, is an exceptional bird regarding the property flying.thing, then such a statement cannot be expressed in normal default logic.

Reiter and Criscuolo, (Reiter, 1981), to overcome this problem, introduced a second form of default called *semi-normal default logic.*

A default is said to be *semi-normal* if it is of the form:

$$\hat{D} \qquad \frac{\alpha(x) : \beta(x) \wedge \gamma(x)}{\gamma(x)}$$

where, in addition to the conclusion $\beta(x)$ being referred, exceptions are explicitly coded in the $\gamma(x)$. $\hat{D}$ is the semi-normal version of the normal default D.

Hence, we may represent Henry the ostrich as an exceptional bird regarding the property flying.thing, using $\hat{D}$ as follows:

$$\frac{Bird(Henry) : \neg Ostrich(Henry) \wedge Can\text{-}fly(Henry)}{Can\text{-}fly(Henry)}$$

Default logic represents facts and rules but not the implicit inference mechanisms behind commonsence reasoning, because (semi-normal) default logic rules need to represent inheritance exceptions explicitly. In addition, semi-normal default logic fails to solve ambiguity because it follows the credulous approach: Default logic cannot by itself say which of two or more extensions is the preferred or intuitively correct one.

To recap, Etherington and Reiter (Etherington and Reiter, 1983) introduced a logical formalization of inheritance. Touretzky (Touretzky, 1984), in response, developed a path-based reasoner using nonstandard logic. His approach criticized Etherington and Reiter's default logic. In the spirit of Etherington and Reiter (Etherington and Reiter, 1983), Etherington (Etherington, 1987b) presented a translation of Touretzky's inheritance network into default logic, using proof theory as minimum correctness criteria for inheritance determination. This is done by adopting an inheritance network with five link types using default and first-order formulae.

Etherington (Etherington, 1987a; Etherington, 1987c; Etherington, 1988) argued that a formal logic system for nonmonotonic reasoning is to be preferred to semantic network approaches. The advantages include increased clarity, provable correctness, and the provision of useful tools for the specification and description of nonmonotonic systems. Etherington criticized Touretzky's claim that adding new information implies changing current defaults and that the translation of a link depends on other links in a network, by stressing that DL can only represent and therefore generate extensions of inheritance structures. Etherington explored the correspondence relation between inferential distance ordering and default logic in that inferential distance ordering can be seen as a restricted default logic. Finally, Etherington (Etherington, 1994) made brief comments on his previous approach (Etherington, 1987b), namely 'formalizing nonmonotonic reasoning systems'. Under a section entitled 'Open Questions', Etherington raised the question of whether specificity, inferential distance ordering, and related ideas could be generalized into arbitrary default theories.

Poole (Poole, 1988; Poole, 1991) presented an informative introduction to the debate concerning the role of classical logic in default reasoning and nonmonotonic reasoning systems. Poole recognized that default logic does not consider some cases in which the expected answer is not produced. The reason is that there is nothing in the definition of extensions that requires them to be consistent with each other and with what is believed. For instance, Poole presented the following example (Brewka, 1991):

$$: usable(x) \wedge \neg broken(x) / usable(x)$$

$$broken(leftarm) \vee broken(rightarm)$$

In this example, although we know that one arm is broken, the default theory has one extension containing both usable(rightarm) and usable(leftarm). This problem concerns the use of a *non-normal default*.

Following the spirit of Israel's (Israel, 1987) approach, and instead of going along with extensions to standard logic, Poole argued that nonmonotonicity is a problem of how logic is used rather than being a problem with logic itself. Poole considered reasoning not as a form of deduction but as theory formation. Poole showed that the idea of theory formation from a fixed set of possible hypotheses is a natural characterization of default reasoning. This kind of logic tells us what the theory predicts rather than what the theory deduces. His approach assumes that the user provides a set of formulae expressing facts and hypotheses that can be used as explanations for nonmonotonic conclusions. A simple semantics comparable with Reiter's default logic was proposed in which Poole discussed the translation features of semi-normal defaults when applied to inheritance problems using semantic networks. A simple programming language for default reasoning (THEORIST) was adopted.

Brewka (Brewka, 1989; Brewka, 1991) accommodated generalized preferred subtheories, a logic that considers Poole's approach of default reasoning (Poole, 1988) as an instance. Brewka asked what makes a default a default, and what the differences are between a default and a fact. The core of Brewka's notion is that of preferred default subtheories, in that the different levels of default theory can be seen as directions to define specific preference ordering on maximally consistent subsets. Brewka provided two generalizations of Poole's system. The first generalization presents several layers of possible hypotheses that represent different degrees of reliability; the second generalization uses partial orderings between premises in order to distinguish between more and less reliable formulae.

3.3.6 Circumscription

McCarthy's (McCarthy, 1980; McCarthy, 1986) notion of circumscription was one of the very first to be proposed for formalizing nonmonotonic reasoning. The idea of circumscription is to say that we do not have to state explicitly the abnormality or the exceptions in one domain in order to minimize the domain's extension. Circumscription allows us to ignore an implicit exception by assuming that it is false unless something is explicitly abnormal. To achieve this, McCarthy augmented first-order logic with a predicate (second order) that expresses a variety of default rules, each with its own abnormality. In addition, and in order to minimize the number of abnormal objects, McCarthy accepted as abnormal only the known abnormal objects: that is, individuals and classes that have been explicitly told to be abnormal. For example, the default 'Cats are mammals' can be represented in the form:

$$\forall x.Cat(x) \wedge \neg AB(x) \supset Mammal(x).$$

which is interpreted as 'cats which are not abnormal are Mammal'. 'AB' here is a predicate that means 'with respect to Mammal but not other aspects'.

In the literature various forms of circumscription are adopted, such as predicate circumscription (McCarthy, 1980) and domain circumscription (McCarthy, 1977). For example, predicate circumscription assumes that just under specific facts, entities will satisfy a given predicate, whereas domain circumscription assumes the known entities are all there. Other forms of circumscription are proposed as well. These include variable circumscription and prioritized circumscription (Konolige, 1989), (Lifschitz, 1986a; Lifschitz, 1986b). Circumscription is a general formalization of nonmonotonic reasoning, many think the most promising one (Brewka, 1987)).

3.3.7 Conditional logic

Conditional logic, adopted by Delgrande (Delgrande, 1987; Delgrande, 1988), is another approach to formalizing and representing default reasoning. Delgrande introduced the conditional logic N for deriving defaults for nonmonotonic inference. Delgrande's logic N consists of first-order logic with a binary connective $\Rightarrow$, as an extension, called the *variable conditional.* The variable conditional N is intended to mean in terms of 'possible worlds' or in terms of 'the world at hand is one of the least exceptional worlds consistent with what is known'. If α and β are formulae, then $\alpha \Rightarrow \beta$ is also a formula with the intuitive meaning 'all other things being equal, if α then β', or 'if α then normally β' (Delgrande, 1988).

Delgrande's conditional logic can be represented by the following:

$$\Diamond\alpha \supset ((\alpha \Rightarrow \beta) \supset \neg(\alpha \Rightarrow \neg\beta)).$$

which can be read as 'if α is possible and $\alpha \Rightarrow \beta$ is true, then it is not the case that $\alpha \Rightarrow \neg\beta$ is true' (Delgrande, 1988). For example (Delgrande, 1988), we may represent the following statement, 'Ravens are normally black, but Albino ravens are normally not black', using logic N as follows:

$$(x)(Raven(x) \Rightarrow Black(x)),$$

$$(x)((Raven(x) \wedge Albino(x)) \Rightarrow \neg Black(x));$$

Delgrande distinguished between reasoning *with* defaults and reasoning *about* defaults. On one hand, reasoning *with* defaults intends to mean the nonmonotonic derivation of inferences and conclusions from defaults (e.g., Reiter's default logic and McDermott and Doyle's modal logic). Delgrande (1988) wrote:

> In Reiter's approach, if we knew that every penguin had to be a bird and that birds normally fly but that penguins do not normally fly, there is no means within the system of concluding that birds that are not penguins normally fly. Similarly, 'penguins are birds' and 'typically penguins are not birds' is never applied in Reiter's system.

On the other hand, reasoning *about* default (Delgrande's approach) is the derivation of defaults from other defaults. Because reasoning about defaults is monotonic in that it derives defaults from other defaults, Delgrande introduced the conditional logic N for deriving defaults for nonmonotonic inference [8].

Delgrande default theory T is an ordered pair $\langle D, C\rangle$ where D is a set of default sentences (or its negations) representing how the world must be, whereas C is a set of contingent facts, that is, how the world being modeled *is*. For example, if T consists of

$$D = \{(Raven(x) \Rightarrow Black(x), (Raven(x) \wedge Albino(x)) \Rightarrow \neg Black(x)),\}$$

$$C = \{Raven(Opus)\},$$

then we can make the default conclusion ρ, that is, '$Black(Opus)$'.

3.3.8 Probabilistic reasoning

Neufeld (Neufeld, 1991), Bacchus (Bacchus, 1989), and Neufeld and Poole (Neufeld and Poole, 1990) adopted a different approach to nonmonotonic reasoning through the use of probabilistic reasoning. The basis of probabilistic reasoning is that there is randomness in the world. Because with some cases it is impossible to estimate or predict the world with certainty, some statistical knowledge about the likelihood of various outcomes will help when dealing with problems of ambiguity and conflict.

After comparing the semantics of typicality assertions and universally quantified assertions, Bacchus (Bacchus, 1989) introduced a style of formalization based on dividing an inference engine into inductive and deductive parts. These parts have the ability to reason semantically with any inheritance structure that combines strict and defeasible assertions. His system is unlike other statistical systems, such as those provided

[8]Conditional logic represents monotonic logic of the form 'if ... then', for example, 'if Clyde is an elephant, then Clyde is a grey.thing'. In order to move a representation from a monotonic (strict) state to a nonmonotonic state we need to introduce an extra consideration to represent something like almost all elephants are grey.thing or normally elephants are grey.thing, and this is what the variable conditional N does.

by Neufeld and Poole (Neufeld and Poole, 1990), who interpret defeasible typicality assertions as unconditional probabilities that are less than conditional (strict) probabilities. This system interprets the defeasible typicality assertions as being statistical assertions having a statistical majority interpretation. The syntax and the semantics for the encoded knowledge are presented.

Neufeld (Neufeld, 1991) identified a probabilistic semantics for inheritance structures with exceptions. Neufeld argued that the literature on inheritance hierarchies offers very little in terms of defining the meaning of any particular link, especially in the meaning of *typically*, which is left to the intuition of the reader: Intuitions can clash. Neufeld introduced a (double arrows) link to represent strict relations such as *Royal.elephant* $\Rightarrow$ *elephant* to mean 'all Royal.elephants are elephants'. In his approach, Neufeld emphasized the two types of domain ambiguity that are caused either by the lack of formal semantics or by inheritance in a domain of reasoning under uncertainty. Neufeld represented probabilistic semantics for inheritance structures with exceptions using conditional probability. In order to represent a statement like 'α is a typical β', Neufeld used the probabilistic semantics of the form p($\alpha|\beta$) > $p(\beta)$ as meaning that observing α increases belief in β or increases our expectation in β. As an example, observing that something is a bird increases our belief that it lays eggs.

Shastri's (Shastri, 1989) evidential reasoning involves finding the most likely hypotheses among finite set of alternative hypotheses. Because AI applications require a reasoner to have the ability to reason with uncertain information, and because probability holds uncertainty as one of its features, probabilistic logic is one of the plausible inference formalizations.

Shastri pointed out that default logic cannot decide which of two or more interpretations of a default inheritance structure should be preferred [9]. Shastri made it clear that his goal was to provide a theoretical framework within which two different ways of opposite polarity of getting to a certain node in a network may be differently weighted, and that the task of an evidential-based theory is to provide ways of describing such differences. He introduced a representational language related to

[9] Etherington (Etherington, 1987b) claimed, perhaps not very persuasively, that this misunderstands the nature and purpose of default logic.

standard semantic network formalisms, and the links are provided with measures that reflect how many instances of a type or token are provided with measures that reflect how many instances of a type or token share values of properties belonging to the type or token. The measures are based on the distribution of values, and these measures are used to provide an ordering graph. Evidence combination rules are then introduced to enable inferencing capability, and definitions are provided of direct and multiple inheritance reasoning.

3.3.9 Other Logical-Based Review Literature

Krishnaprasad and Kifer (Krishnaprasad, 1989) combined logic programming concepts and multi-valued logics in a theory of inheritance. They first distinguished between *proof-theoretic* and *translation semantic* approaches to inheritance networks. *Proof-theoretic semantics* consists of providing algorithms for identifying inheritable paths supported by a network. *Translation semantics* consists of translations of inheritance networks into some standard logical formalism. The work has a strong leaning toward the translation approach. Their starting point is with the term 'more informative than'. For example, knowing that Tweety is a penguin is more informative than knowing that Tweety is a bird because winglessness contributes more evidence in support of Tweety's inability to fly than does bird in support of its flying ability.

Thomason and Horty (Thomason, 1989) provided four-valued autoepistemic logic. Path-based relations are mapped onto expressions in their version and model-theoretic accounts are provided of various properties of inheritance. Thomason and Horty were not convinced that translating inheritance networks into standard nonmonotonic logic is adequate. They identified some problems with Etherington's treatment of credulous inheritance which requires a semi-normal default logic form. Thomason and Horty's approach is bottom-up in that it attempts to link abstract logical theories to implementations for developing multiple levels of theory. Finally, they made it clear that special nonmonotonic logics for inheritance theory are needed that have the expressiveness of a default logic but also have some concepts of priority between them.

Boutilier (Boutilier, 1989) provided a radical alternative approach to proof-theoretic and traditional logic-based model-theoretic semantics for inheritance networks. His approach uses a conditional logic-E which is an extension of that first proposed by Delgrande (Delgrande, 1987; Delgrande, 1988). First, he interpreted links conditionally: For instance A's are normally B's is translated into 'if A holds, then in the normal course of events B holds as well'. A preference relation between models of the network generated by the possible world semantics of conditional logic is specified. This leads to nonmonotonic consequences of a network being those links true in all preference models. After introducing the property of cumulativity (i.e., if a sentence is derivable from some theory then the addition of that sentence to the theory itself should not affect the set of consequences), Boutilier claimed that path-based reasoners do not possess *stability* if the addition of explicit information to information already implicitly derivable has side-effects that make what was previously derivable underivable. Boutilier claimed that calling a link redundant may give the impression that such links add no information to a network, which is not the intention: Rather than calling these links redundant, *independently justified* may be a more appropriate term. 'In general, a link is considered redundant in a network if there exists a set of links that can be construed as the reason for the truth of that link' (Boutilier, 1989).

3.4 Lattice-Based Approach

As explained earlier, some representation languages provide a way of describing the differences between differently weighted links, each with opposite polarity to a certain node, instead of just preferring one over the other (see (Shastri, 1989)). Some other languages concentrate on labeling the degree of belief in a sentence because sentences can be ordered partially by their truth of falsity $\{t, f\}$. Fuzzy logic is an example of these languages. Lattice-based approaches have a different view here. If we can associate with the partial order relation a greatest lower bound (glb) and a least upper bound (lub), then we will get a *lattice* (Ginsberg, 1988), where *glb* of two formula is their conjunction and the *lub* is their disjunction (Thayse, 1988). A lattice is a set $\{L, \bigwedge, \bigvee\}$, where $\bigwedge$ is called the *meet* operation and $\bigvee$ is called the *join* operation of the lattice L. That is, if $l \leq r$, then $l \bigwedge r = glb(l, r) = l$, and $l \bigvee r = lub(l, r) = r$ (Ginsberg, 1988). In other words, lattices concern the idea of labeling sentences as more than just 'true' or 'false'. Figure 3.10 represents a lattice graphically, where f represents the minimal element of the lattice, and t represents the maximal element.

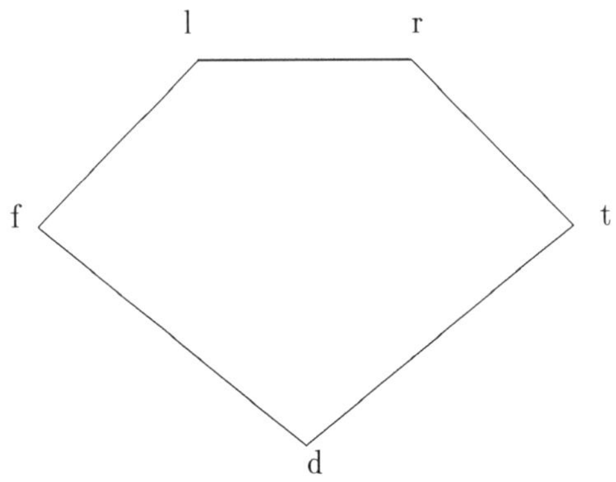

Figure 3.10: A lattice representation.

Lattice-based representations depend mainly on the features of lattice theory. Lattice sets are fully ordered sets, whereas acyclic graphs are partially ordered sets. Lattices are more specific than acyclic; they require a unique meet and a unique join (for more details about lattices see (Ginsberg, 1988), (Thayse, 1988), and (Padgham, 1989)).

As far as inheritance is concerned, Padgham introduced a lattice-based typing model using the notion of types as collections of typical features (Padgham, 1991). Padgham distinguished between two kinds of typing; *prescriptive typing* and *descriptive typing. Prescriptive typing* refers to the process of labeling an instance under a particular type by obtaining certain default properties of the instance (such as in object-oriented languages and inheritance reasoning). *Descriptive typing* refers to the process of classifying an instance as being of a certain type by finding where in a hierarchy a given instance can appropriately be placed (such as in diagnosis and classification tasks). Padgham defined *features* as attribute value pairs such as *weight, 5kg.* She distinguished between two sets of features, namely, *type default D*, which refers to the characteristics in terms of which we think about an object when we use the type name, and *type core* C, which refers to the characteristics that are necessarily required to be present in all instances of the type. These types are represented inside a descriptor called *feature descriptor FD*. FD presents part of all possible subsets of the same universe $\mathcal{C}$. Type default contains more information than type core, that is, all the type core features are necessarily included inside the feature type default. Padgham compared between the core types and the default types, as feature descriptors, using the $\sqsupseteq$ symbol, which means contain more information than. Also she combined them together using $\sqcup$ and $\sqcap$ symbols, which intend to mean union and intersection, respectively, to form a new FD. Padgham claimed that her model has more expressive power in representing inheritance structures than network models. She wrote:

> The model described here has a greater expressive ability than network models for inheritance where the network has only two links types. This is due to the ability to discriminate as to what is definitely inherited, versus what is inherited in typical cases. (p. 66).

Using the lattice-based approach, Padgham et al. highlighted the use and the usefulness of lattice-based as well as negative information in inheritance-based reasoning (Padgham, 1989; Padgham, 1992). Padgham argued that in some domains, such as medical diagnosis (e.g., in order to rule out certain diseases), negative information regarding what an entity is not is important. Padgham adopted a method of default reasoning that allows one to draw negative conclusions that are useful and not available in some other models using negative inheritance reasoning. She provided a basic inheritance model as well as a reasoning method for both positive and explicit negative information, which results in the deduction of more negative information. Padgham distinguished between the strict negative information such as *x is definitely not y*, which is equivalent to *is-not-a* in positive reasoning, and default negative reasoning such as *x is probably not y*. Padgham provided a specific medical diagnosis example to illustrate this new reasoning style (Padgham, 1992).

To sum up, Chapter 2 focused on the nature of inheritance and this chapter presented the various formalisms that can be adopted to capture the nature of nonmonotonicity. The best known logical formalizations, among others, are as follows:

McDermott and Doyle's (McDermott and Doyle, 1980) approach is a consistency-based approach that uses modal operators. Moore's AEL (Moore, 1984; Moore, 1985) and Konolige's HAEL (Konolige, 1988b; Konolige, 1988a) logics are meant to capture the notion of belief and knowing, respectively. Reiter presented default logic (Reiter, 1980) that contains a collection of default rules and different semantics from first-order logic. Reiter and Etherington's (Etherington, 1987a) aim was to model the inheritance specialization principle using default logic. McCarthy's (McCarthy, 1980; McCarthy, 1986) circumscription is meant to capture and use the minimization principle by adding a second-order axiom to first-order theory. Neufeld (Neufeld, 1991) and Bacchus (Bacchus, 1989) among others, tried to capture the notion of nonmonotonicity using a probabilistic approach. Also, from the path-based point of view, different systems are presented. The best known formalizations, among others, are as follows: Fahlman's NETL (Fahlman, 1979), Touretzky's TMOIS (Touretzky, 1986), Horty's skeptical reasoning (Horty et al., 1987), and Stein's Ideally skeptical reasoning (Stein, 1989).

In general, the nonstandard nature of the various systems and their diversity has made it hard to understand them and to compare between them. How are these rules found or justified? Moreover, how can we determine if these rules are reasonable and fit human commonsense reasoning? Also the problem of assigning values to the probabilistic approaches, in particular, is difficult and inadequately understood. As each event within these approaches is a unique occurrence and involves a different prior knowledge state, it is not possible to ask whether the prediction corresponds to actual frequency or not. Also, as explained in Chapter 2, an interpretation procedure in accord with human intuition concerning the meanings of inheritance structures is missing within these formalizations.

In this chapter, the different approaches regarding nonmonotonic reasoning systems are explained, and the following chapter demonstrates the central problem to this book, namely, a clash of intuitions. In Chapter 5, I present a new way to overcome the shortcomings of the already presented formalizations.

Chapter 4

The Problem: A Clash of Intuitions

4.1 Introduction

Touretzky (Touretzky, 1984) introduced a sophisticated mechanism to choose between paths in an inheritance hierarchy: inferential distance ordering (IDO). In formal terms, inferential distance ordering can be explained as follows. An inference path given by:

$$A \rightarrow ... \rightarrow B \not\rightarrow P$$

should pre-empt the path:

$$A \rightarrow ... \rightarrow C \rightarrow P$$

iff there is a path

$$A \rightarrow ... \rightarrow B \rightarrow ... \rightarrow C$$

establishing for A that B is a subclass of C. In other words, subclasses are considered to be more specific than superclasses and the former override the latter. This is referred to as the *specialization principle.*

Among the various pre-emption algorithms that have been proposed, Touretzky (Touretzky, 1986) described a formal mechanism called *on-path pre-emption*, which comes into play when a redundant link causes problems in an inheritance structure. The general idea is that where there is a redundant link, another path can override paths containing the redundant link provided that certain other path considerations are met. The effect is to implement formally the basic requirement that more specific information should override more general information, that is, that information from more specific superclasses should override information from more general superclasses in conflict cases.

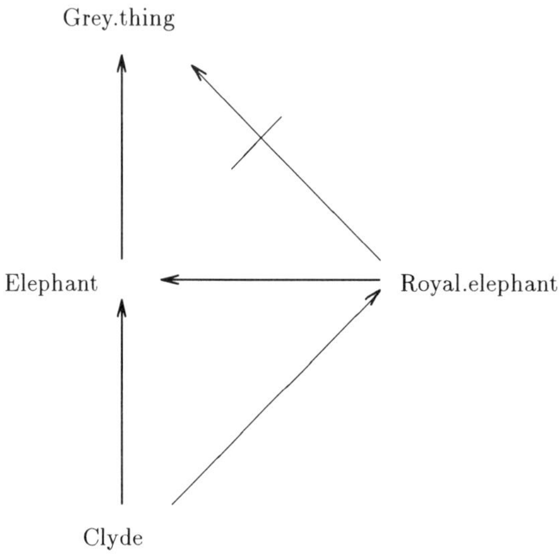

Figure 4.1: Inheritance network with redundancy

Consider Figure 4.1 as an example: The path $Clyde \rightarrow elephant$ is redundant, because this path can be derived from the two primitive paths $Clyde \rightarrow Royal.elephant$ and $Royal.elephant \rightarrow elephant$. According to on-path pre-emption, because there are two paths starting at $Clyde$ of opposite polarity,

$$Clyde \rightarrow elephant \rightarrow Grey.thing$$

and

$$Clyde \rightarrow royal.elephant \not\rightarrow Grey.thing$$

the path that contains as a subpath the path that starts at an on-path intermediary ($Royal.elephant$) overrides the other path starting at Clyde (i.e., Clyde → elephant → Grey.thing). Thus, the notion of on-path intermediary crucially depends on the existence of a redundant link that bypasses that on-path intermediary. However, Sandewall (Sandewall, 1986) argued that on-path pre-emption is too restrictive when dealing with certain classes of inheritance structure. He introduced what Touretzky (Touretzky, 1986) called *off-path* pre-emption, where the idea that more specific, explicit information should win is built into the pre-emption mechanism. Sandewall (Sandewall, 1986) noted that Touretzky's definitions did not take into account cases where the redundant link was itself interrupted by another node (i.e., where a redundant link contains two or more sublinks).

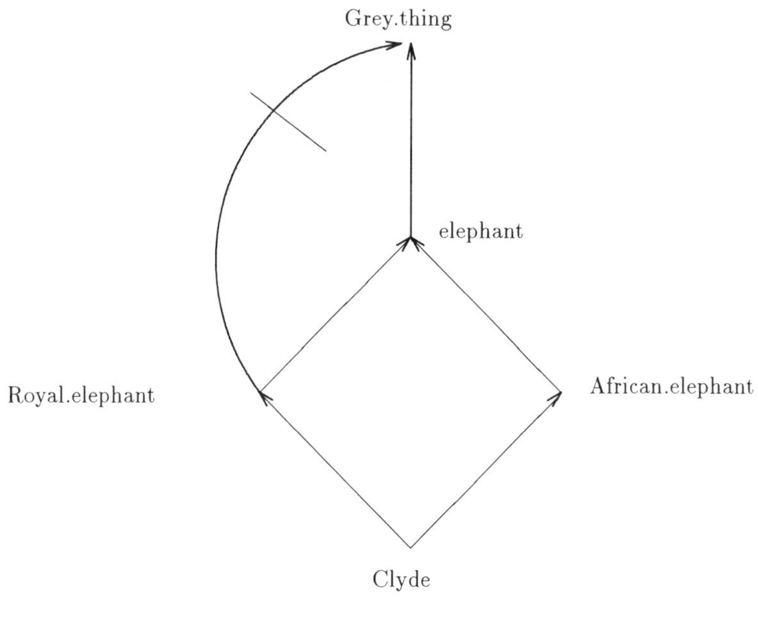

Figure 4.2: Clyde the Royal, African.elephant.

Sandewall (Sandewall, 1986) adopted the example in Figure 4.2 to point out the shortcoming of the on-path approach. The redundant link between *Clyde* and *elephant* is interrupted by the *African.elephant* node, thereby resulting formally in the loss of the redundant link. On-path pre-emption would then not be applicable, and given a skeptical approach nothing could be inferred about Clyde's color. Sandewall's point is that we should nevertheless still let the *Clyde* $\not\rightarrow$ *Grey.thing* conclusion through, because this supports our intuition that *African.elephant* does not affect that part of the network dealing with Clyde's greyness. Because one of Clyde's two immediate classes has some explicit information (i.e., Royal.elephant $\not\rightarrow$ Grey.thing), this should be preferred in cases of conflict. Sandewall's proposal of off-path pre-emption was intended to build this preference explicitly into the inferential mechanism.

However, Touretzky, Horty, and Thomason (Touretzky et al., 1987) argued that Sandewall's off-path is less intuitive than Touretzky's on-path, in that off-path pre-emption also returns counterintuitive results, citing Figure 4.3 as an example of where the analogous intuitive conclusion that George drinks beer is not supported.

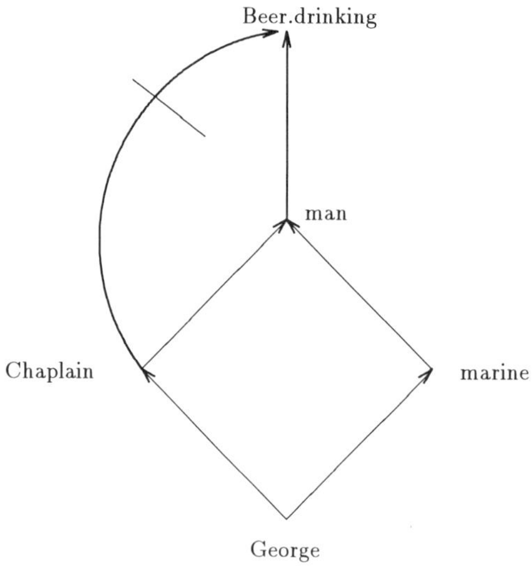

Figure 4.3: George the Marine-Chaplain.

In the George example there is a problem as to whether George does or does not drink beer, with intuition saying that he does. If *drinks.beer* is not an inheritable property, then information about beer drinking must be attached to classes and individuals explicitly. The debate as to whether George is or is not a beer drinker now boils down to where *drinks.beer* should be encapsulated. If within *man*, this is not inheritable by subclasses, and similarly with *marine*, where George cannot inherit uninheritable properties. The 'intuitive' conclusion that George does indeed drink beer can be explained in terms of intuition deciding that beer-drinking is such a strong acquired property for marines that it may be inheritable. However, if it does, we seem to have a case of more general information overriding more specific information (i.e., the information attached to *man* overrides the specific information about *chaplain*.

But the question that we want to ask is, do Figures 4.2 and 4.3 (as mentioned in (Touretzky et al., 1987)) have the same topology? As we stated earlier, the fact that the intuitive reading in the George structure seems to require more general information to override more specific information (i.e., information attached to *man* (*drinks.beer*) to override information attached to *chaplain*) reduces the usefulness of pre-emption as a whole.

4.2 Summary

The problem that sticks with nonmonotonic reasoning systems in general and nonmonotonic inheritance reasoning systems in particular is the lack of a general theory that supplies us with specific rules and definitions that can help in cases of conflict and ambiguity. The implication of the above debate between the proposers of on-path and off-path pre-emption is that proposed solutions seem to depend on human intuitions rather than specific definition, rules, and procedures, and therefore as Touretzky put it (Touretzky et al., 1987), 'a clash of intuitions'.

For instance, there is the question regarding the 'George-marine' example chosen by Touretzky et al. of how the intuitive interpretation is obtained: If more general information is allowed to override more specific information, what are the implications for pre-emption as a whole? How general is pre-emption? What classes of (semantic) problem can pre-emption (both on-path and off-path) deal with, and why? In the final analysis, the formal basis of pre-emption - that more specific information should override more general - may be shown to be not general enough. Whereas Touretzkt et al. (Touretzky et al., 1987) argued that on-path pre-emption is more intuitive than off-path pre-emption, some (e.g., (Etherington, 1987b)) remain neutral, and others (e.g., (Sandewall, 1986)) prefer the more permissive off-path pre-emption.

In this book (the following chapters), I propose a new inheritance theory and a set of procedures and rules for handling nonmonotonic inheritance reasoning. In addition I provide a general algorithm for handling clashes of intuitions, such as the George-Marine problem discussed earlier.

Chapter 5

EIR: An Exception-Based Approach to Nonmonotonic Reasoning

This chapter describes procedures in which exceptions (exceptional classes) can be identified and subsequently located in inheritance structures. I also analyze the on-path versus off-path debate using EIR (Al-Asady, 1993a), (Al-Asady, 1996).

5.1 Introduction

Inheritance is the main inference and knowledge-sharing mechanism in many class-based and object-centered knowledge representation systems (Brachman, 1983); (Patel-Schneider, 1991). A class is a description of a group of objects related through similarity of structure and behavior. Through the process of *specialization*, class information can be reused by creating subclasses that add to or modify the information of the classes to which they belong (their superclasses). Classes are defined in such a way that the total class structure, usually represented by some taxonomic structure and associated *inheritance graph*, typically form a hierarchy: The root class will contain the most general information, and the leaf classes will contain the most specific. In many KRSs, specialization consists of simply adding more information to that already inherited from a higher class (as opposed to modifying or deleting information).

However, when modification and deletion are allowed, *masking* mechanisms must be implemented so that the information in the more specific classes overrides, or hides from view, information that would incorrectly be inherited. For instance, if told that Tweety is a bird, and that birds fly, we infer through normal inheritance that Tweety flies. If then told that Tweety is a penguin, that penguins cannot fly, and that penguins are nevertheless birds, not only must the information that penguins fly be masked (the property of flying would normally be inherited from the class of birds) but so must the information that Tweety flies. Masked properties must be 'inherited' by subclasses also. Through masking, inheritance systems can be used to represent nonmonotonic reasoning: First, Tweety is a bird, and birds fly; second, Tweety is a penguin, and penguins are birds, and penguins do not fly. In terms of classical logic, a system is *monotonic* if whenever a formula p is derivable from a set of premises P then p is also derivable from each superset of P. If p is 'Tweety flies' and P1 is {'Tweety is a bird', 'birds fly'}, then because p is no longer derivable from a superset of P1, namely, P2 = {'Tweety is a bird', 'birds fly', 'Tweety is a penguin', 'penguins are birds', 'penguins don't fly'}, we have *nonmonotonicity.*

5.2 Exceptional Inheritance Reasoning

In this section I propose a new way of representing and handling nonmonotonic inheritance structures. The representation schema I use is called *exceptional inheritance reasoning* (EIR). Exceptional inheritance reasoning depends on distinguishing between (a) *typical* and *exceptional* classes, and (b) *inherited* and *acquired*[1] properties.

5.2.1 Typical and Exceptional Classes

A typical class is one that passes information to its subclasses or instances by default, as is the case with standard class specialization. If the inheritance structure consists solely of typical classes linked by *is-a* links, there is no problem in identifying the typical properties to be inherited by instances. But a class may also be typical by virtue

[1] Not inheritable. See (As-Sadr, 1987).

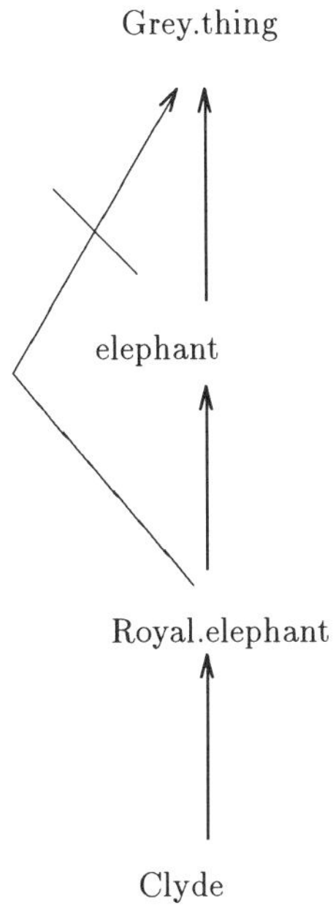

Figure 5.1: The Clyde-Royal.Elephant network.

of providing exceptions to more general typical classes. The approach to be described here allows exceptional information that is nevertheless inheritable to be encapsulated in a class, called an *exceptional class.* Consider the 'Clyde' structure (Figure 5.1). Clyde, although an elephant (through two *is-a* links), is a Royal.elephant, which typically is not a grey.thing. *Elephant* in this example is a typical class in that elephants typically are grey things. *Royal.elephant*, though, contains an exception to the typical class to which it belongs, (*elephant*). EIR's interpretation of this is that *Royal.elephant* must have inherited the property of typical nongreyness from an exceptional class for which nongreyness is typical. Information which is exceptional to the typical class and belongs to the exceptional class must be inherited as typical by subclasses and instances of the exceptional class, unless overridden.

The reasoning behind this interpretation is as follows. In the current literature the *is-a* links are used to specify superclass relationships as well as to infer inheritance relationships in the typical class case. In exceptional cases the *not-isa* link specifies a nonsuperclass relationship but nevertheless requires a property to be inherited, namely the opposite of what would be typically inherited. This is questionable: If *is-a* is used to specify class relationships as well as inheritance, then *not-isa* should be used for explicitly ruling out certain class relationships (therefore stronger than no links at all) and not permitting any inheritance of properties of any sort. However, *not-isa* is usually used to convey the information that the negation of the property referred to by a class name is to be inherited. For instance, in the previous example, the *Royal.elephant not-isa grey.thing* link is to be interpreted as: 'the class of Royal.elephants is not a subclass of grey.thing, *and also* the class of Royal.elephants is to inherit nongreyness so that its own subclasses and instances can inherit this property.' EIR drives a clear conceptual wedge between *is-a* and *not-isa* here: It is the exceptional class that conveys the missing, exceptional inheritable property, not the *not-isa* link. The exceptional class therefore allows the opposite or negation of what would be typically inherited to be inheritable in its own right. This would leave *not-isa* to perform the simple but important task of explicitly ruling out certain subclass-superclass relationships, without necessarily being involved in aspects of inheritance in any way. The role of EIR here is to create an exceptional class that can be inserted into the structure so that the following, new path results:

$$Clyde \rightarrow Royal.elephant \rightarrow \Psi \not\rightarrow grey.thing$$

where Ψ is the exception class that is just like the *elephant* class, except that it is a subclass to a class (non-grey things) that stands for the opposite of what elephants typically are a subclass of (grey things).

5.2.2 Acquired and Inheritable Properties

A second distinction is between *acquired* properties, which are properties that an object picks up from the environment as a result of interaction with the environment (out of consideration when we deal with typical class's properties), and *inheritable* properties, which are properties that an object can inherit from its class(es). Inheritable properties refer to the typical class properties. I need this distinction to establish a theory for inheritance and to explain certain differences between subclass and superclass properties. The underlying rationale is that an inheritable property has the same meaning no matter what or who it is inherited by, whereas an acquired property, because it is acquired by an object or class from the environment, cannot be passed by inheritance to other objects or classes without *assuming* that those objects or classes have also acquired this property from the environment. Since this assumption cannot always be guaranteed to hold, acquired properties can only be attached to objects or classes if it is asserted that they have acquired that property from the environment. An example of an acquired property for an individual object is Polly being a bird that cannot fly because, for example, a wing is damaged. An example of an inheritable property, again for an individual object, is Tweety being a bird that cannot fly because, for example, Tweety is a penguin. An example of an acquired property for a class is lecturers being rich because, for example, their employers decide to raise their salaries five-fold. But the wealth of lecturers is an acquired property, and it cannot be assumed by default that this property will be possessed by all individuals and subclasses of lecturer: Temporary and part-time lecturers may have their salaries cut in half, for instance, or employers decide to withdraw the raise. An individual lecturer is rich only if that lecturer is asserted to be rich. An obvious example of an inheritable property for classes is the class of elephants being mammal and therefore inheriting all the properties associated with being mammal and passing on these properties to its subclasses and individuals. An inheritable property becomes inherited once the inheritance mechanisms specify, through analysis of the inheritance graph, which inheritable properties are to be inherited by subclasses and individuals. Thus inheritable properties are not necessarily inherited, because there may be various redundancy and ambiguity problems to overcome. Mechanisms for locating Ψ in inheritance structures are explained in the following section.

5.3 The Exceptional Class, Ψ

Consider the following inference path:-

A → C → P

where P is a property, C is a typical class, and A is a subclass.

This inference path represents the sentence that 'an individual A could inherit the property P via the class C'.

Later, assume that the following extension is added to this inference path: A $\not\to$ P.

In addition to the need for defining the mechanisms that we use to retract or keep our previous conclusion, a clear explanation of how such a conflict arises is needed in the first place. Logic-based approaches are not of great help here because we will have two inconsistent paths with no preference between them.

From a path-based point view, because we do not know via which node A is related negatively to P (because the class C is related positively to P), then there must be another node apart from C that relates A negatively to P. Let us represent this node by the symbol Ψ.

Ψ refers to an unspecified collection of properties. That is, it is not a typical class of default properties; rather, it is a class of similar properties. It is a class where most of its properties have been canceled and what remains are properties similar somehow to the properties in A but not in C.

The path

A $\to$ ... $\to$ C $\to$ P

means that A inherits the property P implicitly via the typical class C, whereas

A $\to$ ... $\to$ $\Psi \not\to$ P

means that A is affected by the explicit properties that exist in Ψ, because we do not know what the properties are that Ψ has by default.

Because A is related by an implicit and positive relation to P via C and negatively via the class Ψ, and because we know explicitly that A not-isa P, then Ψ is a representation symbol of the properties that do not exist in C and that appear in A. Therefore, the extension that goes through this node must be the path that we need to follow. Thus, Ψ is the exceptional class that is just like a typical class except that it stands for the opposite of what a typical class, typically, possesses (by default).

Thus, our network extension is the path:-

A $\to \Psi \not\to$ P.

Figure 5.2 is a structure represented using EIR.

In summary, in multiple inheritance with exceptions, complex class

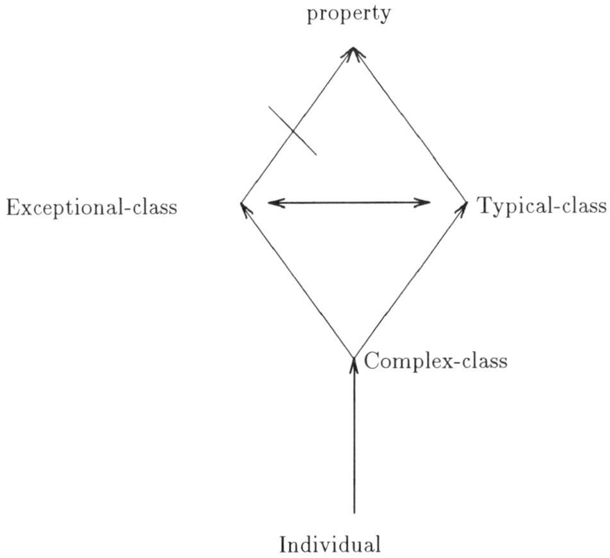

Figure 5.2: Exceptional inheritance structure.

(such as Royal.elephant) properties cannot be explained without the role of the exceptional class Ψ in which all its atypical properties will go through the node Ψ.

5.4 Conceptual Foundations of EIR

I now provide a series of definitions that will make the assumptions and motives behind EIR clear. More detail to support these definitions is provided later in the next section.

Let Ω be a directed, acyclic graph (graph with no loops) containing an inheritance structure Γ. Γ contains three types of 'links' and two types of 'nodes'. The link types are the inheritance positive link *is-a*, the inheritance negative link *not-isa*, and the bidirectional relationship ($\leftrightarrow$), which signifies an opposite relation between a typical class and an exceptional class, both of which are immediate superclasses of an individual.

The node types are as follows, 'instance' or 'individual', which is the first element of an inheritance structure connected by a positive link to its ancestor node and about which we want to infer, such as 'John', 'Clyde', 'Tweety' and so on. We refer to it with the symbol x. The other type is the node 'class' which refers to the type kind, or a collection of similar objects or properties, such as 'elephants', 'birds', 'flying things', 'intelligence things', and so on. I refer to these classes with the symbols α, δ and ζ. I represent these nodes and links in a relation such as

$$x \rightarrow \alpha \rightarrow \delta$$

which means that an instance, such as x, is related by a positive link to the class of type α, in which case it will inherit all the α's type properties and, by default, δ's, whereas

$$x \rightarrow \zeta \not\rightarrow \alpha$$

means that the individual x has a positive relation with the type ζ, but the type ζ has a negative relation (connected with a negative link) with its ancestor α. In the latter case no properties from this type can be inherited by the instance x.

According to this and in order to establish an inheritance mechanism for a directed acyclic graph, I distinguish between two types of (positive and negative) links; strict links, which allow no exceptions to their structure and which do not refer to any typicality, and defeasible links, which allow what the strict links do not.

A *path* is a sequence of related links. A *positive path* is a path in which all links are positive. A *negative path* is a path in which the last link is negative. The set of finite path sequence over Γ is represented by the symbol λ. For example, we represent the assertion 'John is-a human' by the following relation:

$$John \rightarrow human.$$

Now if the class 'human' has the following relation:

$$human \rightarrow has.two.legs,$$

John will inherit the property (has.two.legs) from the positive link between the type 'human' and the type 'has.two.legs'. Similarly, John does not inherit, for example, the property 'has.tail' because the type 'human' has a negative relation with the type 'has.tail' and there is no way for this property to be attached to the instance 'John':

$$John \rightarrow human \not\rightarrow has.tail$$

That is,

$$John \not\rightarrow has.tail.$$

I refer with the symbol η to the *network extensions*, which in turn refer to the set of links that is obtained by applying our reasoning mechanism on an inheritance network. For example, and according to the previous examples, our extension set will be as follows:

$\eta = \langle$ John $\rightarrow$ has.two.legs, John $\not\rightarrow$ has.tail $\rangle$.

I redefine what we mean by multiple inheritance in the light of these comments.

Definition 7 *Multiple Inheritance is the heredity of one or more properties by an individual or by a class, from two or more classes, either of which can be typical or exceptional.*

That is, we broaden the concept of inheritance to take into account the inheritance of exceptional information. However, we need to be more precise about what is inheritable.

Definition 8 *An inheritable property is a property that may pass via a class to a subclass or individual.*

Definition 9 *An acquired property is a property derived from experience and training or as a result of interaction with the environment. This kind of property cannot be inherited by subclasses or individuals of a class that has an acquired property.*

Given these preliminary definitions, 'inconsistencies' in inheritance structures fall into one of the following three categories:

1. The 'inconsistency' is between an acquired property belonging to

one class and an acquired property belonging to another class, where the two properties are negative and positive aspects of one underlying property. Becasue acquired property (see section 5.2.2) is not inheritable, the individual or subclass should inherit neither of the acquired properties. There should be no inconsistency in this case.

2. The 'inconsistency' is between an inherited property belonging to one class and an acquired property belonging to another class, where the two properties are negative and positive aspects of one underlying property. Because an acquired property is more explicit than an inheritable (implicit) property, the individual or subclass should have the acquired property assuming that this individual or class has also acquired this property from the environment. There should be no inconsistency in this case also.

3. The inconsistency is between an inheritable property belonging to one class and an inheritable property belonging to another class, where the two properties are negative and positive aspects of one underlying property. This is the problem case, and EIR is based on the idea that, if one of the classes has the property δ and the other $\neg\delta$, then one class or the other, let us say the one that has $\neg\delta$, must itself have a superclass through which it acquires $\neg\delta$. The explicit insertion of this superclass should help resolve the inconsistency by rerepresenting the inheritance structure in such a way that the relationship between the problem class or individual and its classes and superclasses is disambiguated.
Our concern here is with the second and third types of inconsistency given. Consider the example in Figure 5.1 again:

Clyde is-a Royal.elephant,

Royal.elephant not-isa grey.thing,

Royal.elephant is-a elephant,

Clyde is-a elephant,

elephant is-a grey.thing.

Because *Clyde is-a grey.thing* through the superclass *elephant*, and *Clyde not-isa grey.thing* through a subclass of *elephant*, namely *Royal.elephant*, it follows that there is an inconsistency between these classes regarding Clyde's greyness. To formally solve such an inconsistency so that the result is intuitively similar to the one humans would arrive at (that Clyde is not grey.thing) I need to show that, provided that greyness and nongreyness are not an acquired property, there is some reason for Clyde inheriting one aspect rather than the other. The traditional solution is to use pre-emption, which is to say that Clyde will inherit from whichever class is more specific, where 'specific' is unpacked in terms of what paths exist in the inheritance graph. But, as we shortly see, it is not clear that greater specificity by itself leads to the intuitively correct results (Sandewall, 1986) ; (Touretzky et al., 1987). In the given example, the typical superclass *elephant* cannot explain the property of nongreyness of one of its subclasses, whereas the insertion of an exceptional class that, although similar in every other respect to elephant, nevertheless has nongreyness as one of its properties and is a superclass to the problem class, can. That is, Clyde is not grey.thing because he belongs to a class, *Royal.elephant*, which inherits the property nongreyness from a superclass, which I call Ψ, which is similar to the elephant class except that it possesses the property of nongreyness. This is to be contrasted with the traditional, pre-emptive answer that Clyde is not grey.thing because he belongs to a class, *Royal.elephant*, which on the one hand is a subclass of elephant, itself a subclass of grey things, and on the other hand is a subclass of nongrey things directly. Because the class of Royal.elephant is more specific to Clyde than elephant, Clyde inherits the property of nongreyness. Our claim is that the interpretation based on nongreyness being a property of an elephant-related class to which Clyde the Royal.elephant belongs is to be preferred to the interpretation that because Royal.elephants are not a subclass of grey things and because Clyde is a Royal.elephant, therefore Clyde is not a member of the class of grey things. The problem with the latter interpretation is that it does not always produce intuitively correct results. For example, George is a chaplain who is also in the marines, and marines drink beer but chaplains do not. The intuitive result that George drinks beer arises from information associated with the more general class, in this case marines drinking beer, overriding the more specific information about chaplains not drinking beer. Our interpretation of this example would be that George drinks beer because he belongs to a class that is just like chaplain except for one property of this class is that members of this class drink beer, a beer-drinking chaplain group if you like. This beer-drinking chaplain group is what I signify by Ψ.

In this approach I present a different interpretation of the inconsistency problem that default logic (Reiter, 1980) and inferential distance ordering (Touretzky, 1986) addressed, namely conflict and redundancy problems. Our view here is related to what people mean by *later* information that invalidates a specific property for an individual or a class. I start by asking a specific question: What is the difference between what we know regarding an individual before, and what we are told later? Certainly, it is either the more *specific* information that invalidates the general information because not all instances and individuals share all the features of the general information, or it is the more *explicit* information rather than the default information that overcomes previously derived typicality information. In either case, we have *later information* that invalidates our previous decision in that it forces us to change our mind.

We now provide a definition of exceptional class.

Definition 10 *An exceptional class (called Ψ) for an individual or class X is a typical superclass (if X is a class) or typical class (if X is an individual) created in order to explain inconsistencies (involving inheritable properties only) arising from multiple inheritance with exceptions. It is similar in all respects except one to another typical superclass or typical class (let us say Y) of X. That is, Ψ carries with it everything that Y does except the property in respect of which Y does not. Given Ψ and Y, X will inherit from Ψ rather than from Y.*

This class Ψ has the ability to transfer properties to its related subclasses (complex classes) which may be defined as follows:

Definition 11 *A complex class (called ζ) is a subclass created through the combination of two classes; a typical class as well as an exceptional class Ψ. All its properties are the default (implicit) properties of the typical class unless there is a contradiction in which it will be obtained through the exceptional class Ψ.*

5.5 A Semi-formal Introduction to EIR

Overall, this is what we want our formal schema to achieve. If we have the following inheritance paths:

$$A \rightarrow ... \rightarrow C... \rightarrow P$$

and

$$A \rightarrow ... \rightarrow B... \not\rightarrow P$$

where B is an *immediate* subclass of C but C does not appear in the subpath between B and P, then we want

$$A \rightarrow ... \rightarrow B \rightarrow \Psi \not\rightarrow P$$

where Ψ signifies an exceptional class that is similar in all but one respect to C, namely in respect to $\Psi \not\rightarrow P$. The question of the relationship between Ψ and B's immediate superclass, C, is an open one, because they are exactly the same except for having different aspects of some inheritable properties. We call this open relationship 'bidirectional'.

EIR is all about how we achieve these results. The definitions here are simplified in terms of single superclasses and single inheritable and acquired properties.

Let

- x represent an object.
- α represent x's typical class or superclass that has an *inheritable* property.
- δ be the inheritable property of α.
- ζ be a complex class that is just like α except for the inheritable property with which it conflicts.
- σ is the property in respect to which α and ζ differ (i.e., the exceptional property).
- β be x's class or superclass that has an *acquired* property (β is called 'acquired class' in the definitions that follow).

- γ stand for the conjunction of $\neg\sigma(x)$ and $\neg\delta(x)$ (neither exceptional nor typical inheritable).
- $\odot$ stand for 'provided that'.
- $\otimes$ signify an inconsistency or conflict between two classes with respect to the inheritable property.
- $\triangleright$ stand for a *class similarity* relationship between two classes of an object, *except for* one property (σ) on which they differ.
- $\bowtie$ stand for a *bidirectional* relationship between Ψ and α, both of which are immediate (super)classes of x.

The general form of an inheritance rule in EIR is as follows:

$$\Gamma \odot \Delta \rightarrow \Theta$$

where Γ, Δ, and Θ are sets of formulae that stand for: 'if Γ can be derived and provided that the exceptions Δ have been tested for, then derive Θ.' There are also rules of the form

$$\odot\Delta \rightarrow \Theta$$

where inheritance conclusions require only the exceptions to be tested for. The exception Δ consists of a σ (an exceptional property) or a γ (an acquired property), or both. If an element of Δ is positive, the interpretation is: 'provided that the exception holds', and if an element of Δ is negative: 'provided that the exception doesn't hold'.

Three primitive relations (where $\phi(x)$ has the interpretation of class or object x belonging to ϕ where ϕ stands for a class) are now described. First:

$$PR1 \qquad \zeta(x) \triangleright \alpha(x)$$

that is, x's exceptional class (if there is one) is exactly like a typical superclass α of x except for an exceptional property σ in respect to which α and ζ differ (Definition 4). PR1 can be explained using the bidirectional relation $\Psi \bowtie \alpha$. Then,

$$PR2 \qquad \alpha(x) \odot \neg\zeta(x) \rightarrow \delta(x)$$

and

$$PR3 \qquad \odot\,\zeta(x) \not\rightarrow \delta(x)$$

that is, if x belongs to a typical class and provided that x does not belong to an exceptional class, then x inherits the inheritable property δ of typical class α ($PR2$); and provided that x belongs to an exceptional class, then x does not have the typical inheritable property δ of α ($PR3$).

An *exception indicator rule* is represented as follows (bearing in mind that ζ stands for a class that contains the negation of some inheritable property δ):

$$EIR1a \qquad \alpha(x) \odot \zeta(x) \rightarrow \otimes\delta(x)$$

The following definition describes this rule:

Definition 12 *If x is an α and provided that it belongs to an exceptional class ζ (where $\zeta(x) \rhd \alpha(x)$ by $PR1$), then there is a conflict between these classes regarding the typical inheritable property δ.*

Similarly, an *acquired property* conflict is defined as follows:

$$EIR1b \qquad \alpha(x) \odot \beta(x) \rightarrow \otimes\delta(x)$$

where the conflict concerns whether a typical inheritable property or an acquired property should be inherited. To draw the right inheritance conclusion:

Definition 13 *If x has typical superclass α, and provided that x is neither an instance of an exceptional class ζ nor an instance of an acquired class β, then x inherits property δ.*

This definition is formalized by:

$$EIR2 \qquad \alpha(x) \odot [\neg\zeta(x) \wedge \neg\beta(x)] \rightarrow \delta(x).$$

This leads to the following exception rules:

Definition 14 *If x has typical superclass α, and provided that x is an instance of an exceptional class ζ but not an instance of the acquired class β, then x inherits the exceptional property σ.*

The rule can be represented as follows:

$$EIR3 \qquad \alpha(x) \odot [\zeta(x) \wedge \neg\beta(x)] \rightarrow \sigma(x),$$

Definition 15 *If x has typical superclass α, and provided that x is an instance of the acquired class β but not an instance of the exceptional class ζ, then x has the acquired property γ.*

This definition can be represented as follows:

$$EIR4 \qquad \alpha(x) \odot [\beta(x) \wedge \neg\zeta(x)] \rightarrow \gamma(x).$$

Then,

Definition 16 *If x has typical superclass α, and provided that x is an instance of an exceptional class ζ and also an instance of the acquired class β, then x has the acquired property by assertion.*

This definition can be represented in a rule as follows:

$$EIR5 \qquad \alpha(x) \odot [\beta(x) \wedge \zeta(x)] \rightarrow \gamma(x)$$

In other words, the acquired property overrides an inheritable property. This is because the acquired property is an explicit property (declared and asserted by the user), whereas the inheritable property is a typical (implicit) one.

Finally,

Definition 17 *If x has immediate (super)classes Ψ and α, then $\Psi \bowtie \alpha$, i.e. Ψ and α are bidirectionally related in that they conflict over δ. The bidirectional link cannot be used for drawing inheritance conclusions.*

The above definition can be represented in a rule as follows:

$$EIR6 \qquad \alpha(x) \odot \Psi(x) \rightarrow \alpha \bowtie \Psi$$

Bidirectionality (signified by $\leftrightarrow$ in the following networks) is meant to express the possibility of mutual specialization between a typical class and its related exceptional class. That is, just as an exceptional class is similar to a typical class except for an exceptional property, a typical class can be regarded as similar to an exceptional class except for a typical property.

5.6 EIR Algorithm

In order to specify the EIR algorithm, we refer to the notation described in Section 5.5 and Figures 5.3 and 5.4. Let Ω be a directed acyclic graph containing the inheritance structure Γ. Let x be an object or individual related via a set of positive links (path) (x_1, x_2, ..., x_n) except the last link, namely (x_n,P) may contain a positive or negative relation.

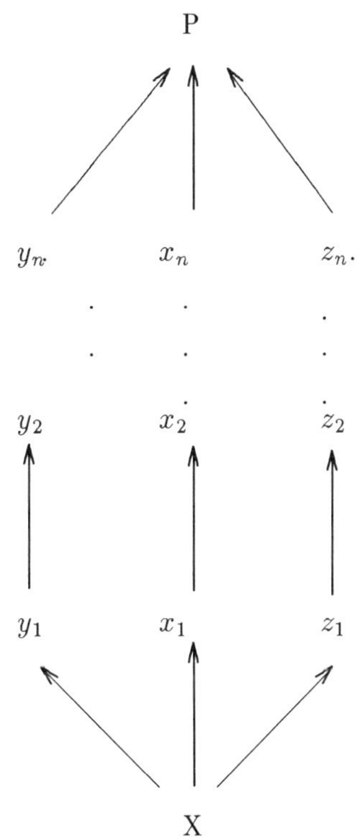

Figure 5.3: EIR hierarchy.

In this algorithm, we refer to typical properties by Assertion 1, $(x_1, ..., x_n)$, exceptional properties by Assertion 2, $(y_1, ..., y_n)$ and acquired properties by Assertion 3, ($z_1, ..., z_n$ as follows: Read the input assertion (i=1). Apply EIR2 on this path to obtain the default relation x $\rightarrow$ P or x $\not\rightarrow$ P, providing that x is not an exceptional class and does not have acquired properties.

$$EIR2 \qquad \alpha(x) \odot [\neg\zeta(x) \wedge \neg\beta(x)] \rightarrow \delta(x).$$

Read the *later* input assertion (i=2) that contains the path (x, y_1, y_2, ..., y_n, ($\neg$) P). If there is no link between x_i and y_i then stop and write 'ambiguous case'. Otherwise (there is at least one link between x_i and y_i) proceed.

If $y_i \triangleright x_i$ (i.e. $y_i \subset x_i$) then check if the path y_1, y_2, ..., y_n is an exceptional path by comparing the specific x_i with y_i. If no contradiction ($\otimes$) exists between the path y_1, y_2, ..., y_n and the path x_1, x_2, ..., x_n then go to Procedure 3. Otherwise apply Procedure (EI) and go to Procedure 4.

Procedure 3: Read the input assertion (i=3) that contains the path (x, z_1, z_2, ..., z_n, ($\neg$) P). Compare z_i with x_i. If no contradiction ($\otimes$) exists between the path x_1, x_2, ..., x_n and the path z_1, z_2, ..., z_n then write that the final extension is assertion number (i=1). Stop. Otherwise apply Procedure (EI). Write that the final extension is assertion number (i=3). Stop.

Procedure 4: Read the input assertion (i=3) that contains the path (x, z_1, z_2, ..., z_n, ($\neg$) P). Compare y_i with z_i. If no contradiction ($\otimes$) exists between the path x_1, x_2, ..., x_n and the path z_1, z_2, ..., z_n then write that the final extension is assertion number (i=2). Stop.

Procedure EI: Locate Ψ as a typical class between y_i and x_i in the exceptional class case, or between z_i and x_i with the acquired properties case.

Present the relation (Ψ,Typical-class(x)) as follows: $\Psi \bowtie$ Typical-class(x).

Add the path x $\rightarrow x_i \rightarrow \Psi$ ($\not\rightarrow$) P to the inheritance structure Γ.

Obtain according to the class type one of either of the final relation

$$Typical.class(x) \odot [\zeta(x) \wedge \neg\beta(x)] \rightarrow \sigma(x).$$

or

$$Typical.class(x) \odot [\beta(x) \wedge \neg\zeta(x)] \rightarrow \gamma(x).$$

We can demonstrate these procedures by the structure in Figure 5.4.

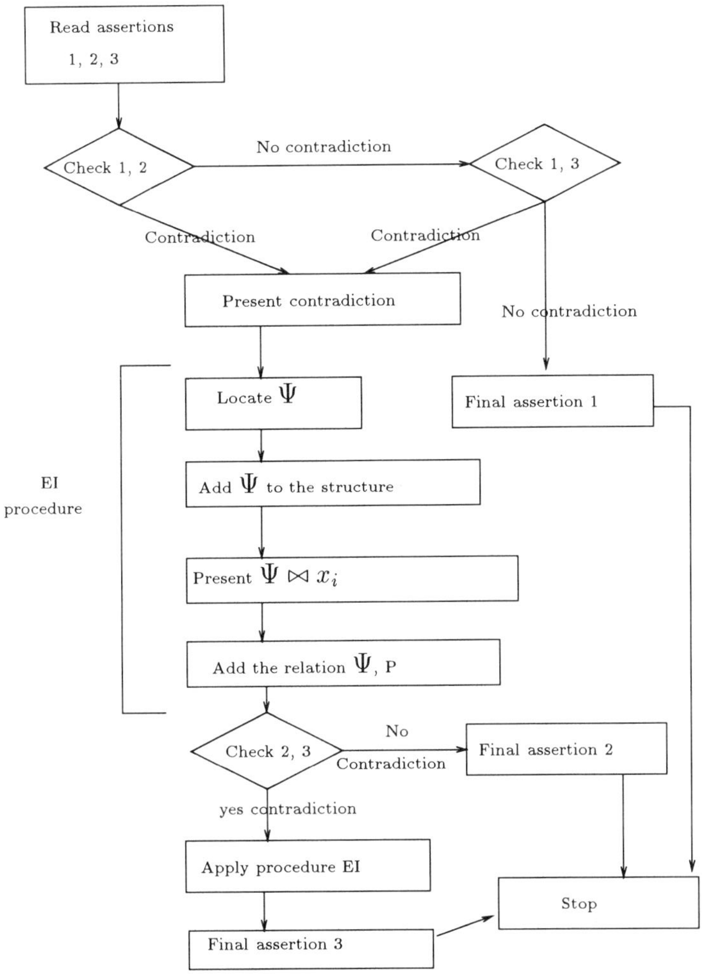

Figure 5.4: EIR algorithm.

5.7 Examples

5.7.1 The Royal.Elephant Problem

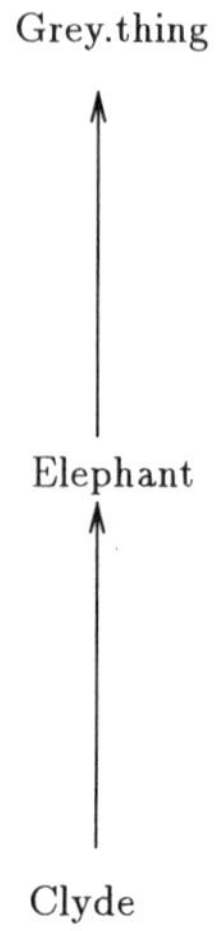

Figure 5.5: Clyde, the elephant.

Consider the following primitive paths (represented in Figure 5.5):

$$Clyde \rightarrow elephant,$$
$$elephant \rightarrow grey.thing.$$

To identify Clyde's default property, *grey.thing*, we need to represent explicitly that *Clyde* is neither a member of an exceptional class ζ nor a member of a class β that has an acquired property, using *EIR2*:

$$Elephant(Clyde) \odot [\neg\zeta(Clyde) \wedge \neg\beta(Clyde)] \rightarrow grey.thing(Clyde).$$

If the following primitive paths are included later (represented in Figure 5.6):

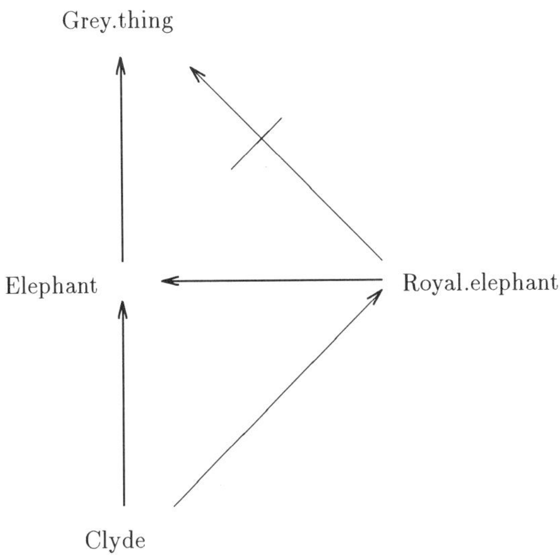

Figure 5.6: Clyde, the Royal.Elephant.

$$Clyde \rightarrow Royal.elephant,$$
$$Royal.elephant \not\rightarrow grey.thing,$$
$$Royal.elephant \rightarrow elephant,$$

we first check whether there is any contradiction with the previous assertion by using PR2 and PR3:

$$Elephant(Clyde) \odot \neg Royal.elephant(Clyde) \rightarrow grey.thing(Clyde).$$

$$\odot Royal.elephant(Clyde) \rightarrow \neg grey.thing(Clyde).$$

These primitive relations identify the contradiction:

$$Elephant(Clyde) \odot Royal.elephant(Clyde) \rightarrow \otimes grey.thing(Clyde).$$

The exception $\odot Royal.elephant(Clyde)$ and the default conclusion $\otimes grey.thing(Clyde)$ in the contradiction definition are used to locate Ψ, namely an exceptional, immediate superclass of *Royal.elephant* that can help us explain Clyde's property. We add:

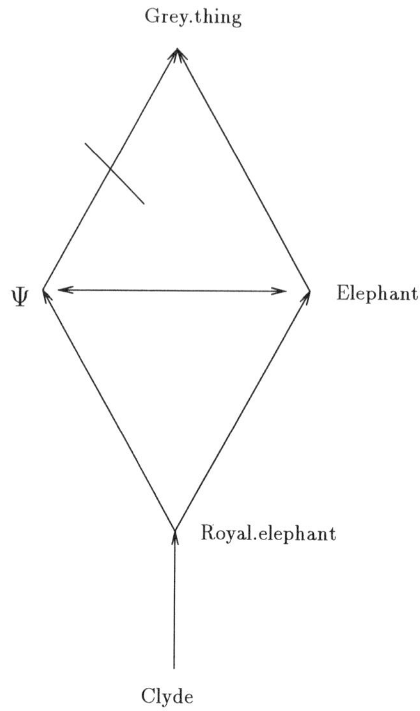

Figure 5.7: The Clyde structure represented using EIR.

$$Royal.elephant \rightarrow \Psi,$$
$$\Psi \not\rightarrow grey.thing.$$

(See Figure 5.7 for the EIR representation of the structure). Ψ is the exceptional class that has the missing, complementary aspect regarding the typical superclass property, and its negative link with that property, namely, $\Psi \not\rightarrow grey.thing$, signifies this. Because *elephant* and Ψ have an opposite relation regarding the individual *Clyde*, it follows by *EIR6*:

$$\Psi \bowtie elephant.$$

We now have the path:

$$Clyde \rightarrow Royal.elephant \rightarrow \Psi \not\rightarrow grey.thing.$$

Using *EIR3* we obtain:

$$Elephant(Clyde) \odot [Royal.elephant(Clyde) \wedge \neg\beta(Clyde)] \rightarrow \neg grey.thing(Clyde).$$

Because there are no acquired properties to consider, the default inheritance conclusion is $\neg grey.thing(Clyde)$.

5.7.2 Clyde, the Three.Legged.Thing

We can also represent new information regarding *Clyde*'s acquired properties. Assume that we are informed explicitly that Clyde is a three.legged elephant. Because three.legged.elephant is a property Clyde picks from the environment, the three.legged property is an acquired property (it is not one of the properties that we could find in a knowledge base regarding inheritable properties).

Consider the following primitive paths (represented in Figure 5.8) :

$$Clyde \rightarrow elephant,$$
$$elephant \rightarrow four.legged.thing.$$

To identify Clyde's default property, *four.legged.thing*, we need to represent explicitly that *Clyde* is neither a member of an exceptional class ζ nor a member of a class β that has an acquired property, using *EIR2*:

$$Elephant(Clyde) \odot [\neg\zeta(Clyde) \wedge \neg\beta(Clyde)] \rightarrow four.legged(Clyde).$$

If the following primitive paths are included later (represented in Figure 5.9) :

$$Clyde \rightarrow three.legged.elephant,$$
$$three.legged.elephant \rightarrow elephant,$$

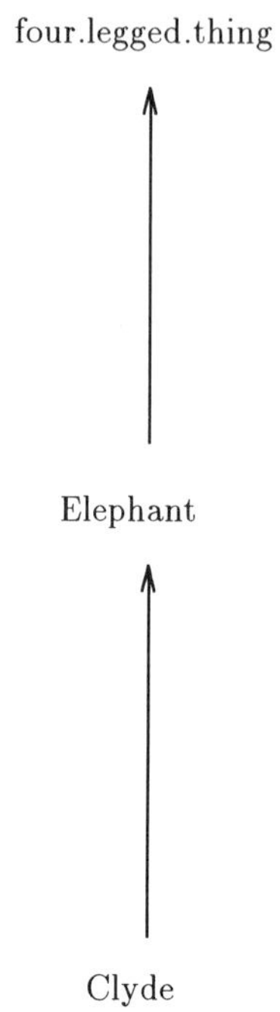

Figure 5.8: Clyde, the four.legged.elephant

we first check whether there is any contradiction with the previous assertion by using PR2 and PR3:

$$Elephant(Clyde) \odot \neg three.legged(Clyde) \rightarrow four.legged.thing(Clyde).$$

$$\odot three.legged(Clyde) \rightarrow \neg four.legged(Clyde).$$

These primitive relations identify the contradiction:

$$Elephant(Clyde) \odot three.legged(Clyde) \rightarrow \otimes four.legged(Clyde).$$

The explicit exception $\odot three.legged(Clyde)$ and the default conclusion $\otimes four.legged(Clyde)$ in the contradiction definition are used to locate Ψ, namely an exceptional, immediate superclass of *three.legged.elephant* that can help us explain Clyde's property. We add:

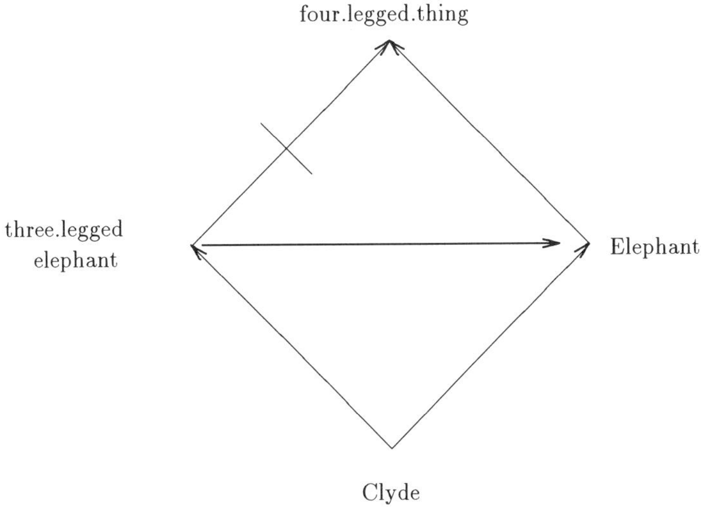

Figure 5.9: Clyde, the three.legged.elephant.

$$\begin{gathered} three.legged.elephant \rightarrow \Psi, \\ \Psi \not\rightarrow four.legged.elephant. \end{gathered}$$

(See Figure 5.10 for the EIR representation of the structure.) Ψ is the exceptional class that has the missing, complementary aspect regarding the typical superclass property, and its negative link with that property, namely, $\Psi \not\rightarrow four.legged.thing$, signifies this. Because *elephant* and Ψ have an opposite relation regarding the individual *Clyde*, it follows by *EIR6*:

$$\Psi \bowtie elephant.$$

We now have the path:

$$Clyde \rightarrow three.legged.elephant \rightarrow \Psi \not\rightarrow four.legged.elephant.$$

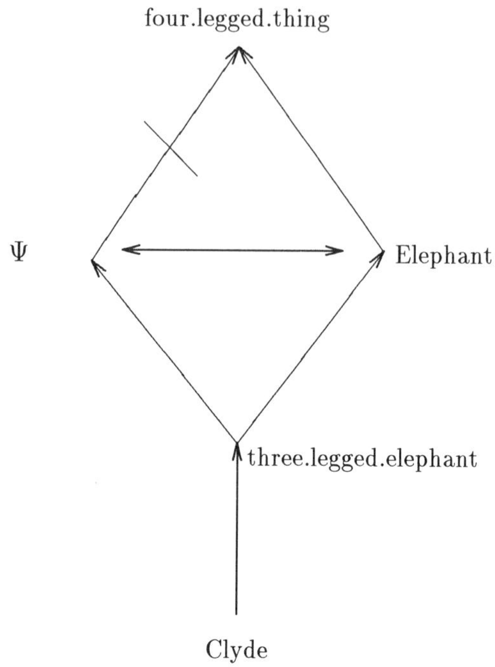

Figure 5.10: The three.legged.elephant structure using EIR.

Using *EIR4* we can represent the acquired property *three.legged.elephant* as follows:

$$Elephant(Clyde) \odot [\neg\zeta(Clyde) \wedge three.legged.elephant(Clyde)] \rightarrow$$
$$three.legged.elephant(Clyde).$$

5.7.3 The Tweety-Penguin Problem

Consider the following paths (represented in Figure 5.11):

$$Tweety \rightarrow bird,$$
$$bird \rightarrow fly.$$

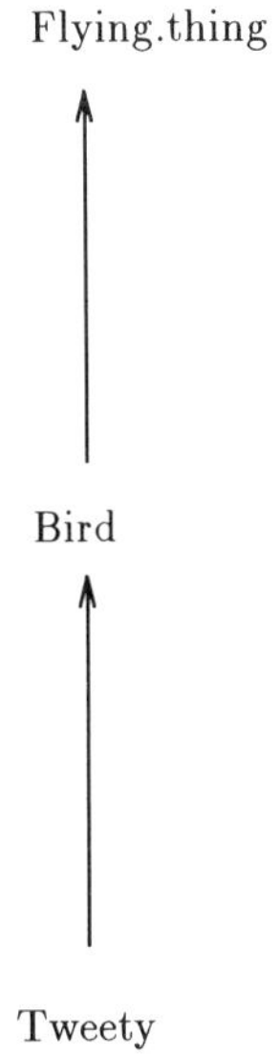

Figure 5.11: Tweety, the bird.

To identify Tweety's default property, *Fly*, we need to represent explicitly that *Tweety* is not an exceptional class and the *Fly* property is not an acquired property, using *EIR2*:

$$Bird(Tweety) \odot [\neg\zeta(Tweety) \wedge \neg\beta(Tweety)] \rightarrow fly(Tweety).$$

If the following primitive paths are included later (as represented in Figure 5.12).

$$\begin{gathered} Tweety \rightarrow Penguin, \\ Penguin \rightarrow bird, \\ Penguin \not\rightarrow fly \end{gathered}$$

we first find out whether there is a contradiction using *PR2* and *PR3* as follows:

$$Bird(Tweety) \odot \neg Penguin(Tweety) \rightarrow fly(Tweety).$$

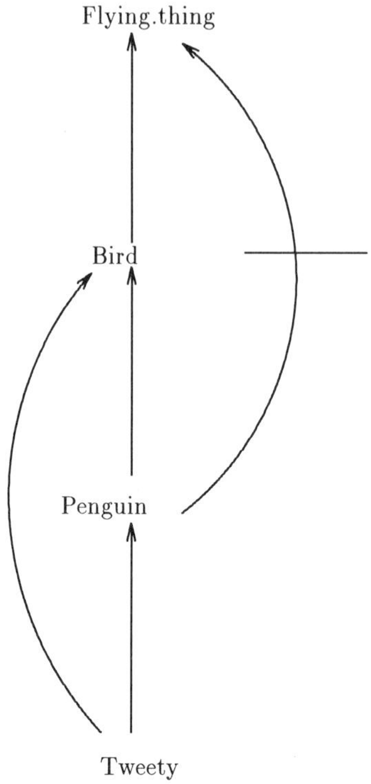

Figure 5.12: Tweety, the Penguin problem

$$\odot Penguin(Tweety) \rightarrow \neg fly(Tweety).$$

Using *EIR1a* we can see that there is a contradiction:

$$Bird(Tweety) \odot Penguin(Tweety) \rightarrow \otimes fly(Tweety).$$

To solve this contradiction we locate Ψ with the following assertions:

$$Penguin \rightarrow \Psi,\ \Psi \not\rightarrow fly,\ \Psi \bowtie bird.$$

In order to associate the property of nonflying with *Tweety*, we need to find another path from Tweety to nonflying (the final set of assertion is represented in Figure 5.13):

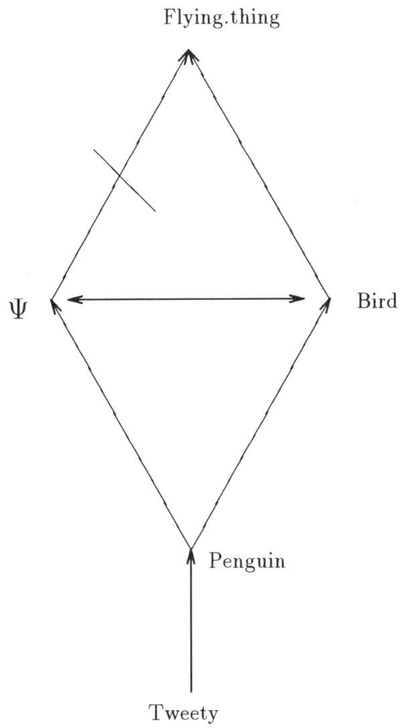

Figure 5.13: The Tweety structure represented using EIR.

$$Tweety \rightarrow Penguin \rightarrow \Psi \not\rightarrow fly.$$

Using *EIR3* we can represent Tweety's property as:

$$Bird(Tweety) \odot [Penguin(Tweety) \wedge \neg\beta(Tweety)] \rightarrow \neg fly(Tweety).$$

The default inheritance conclusion that $\neg fly(Tweety)$ is derived.

5.7.4 The Unicorn Problem

Consider the following primitive paths (represented in Figure 5.14):

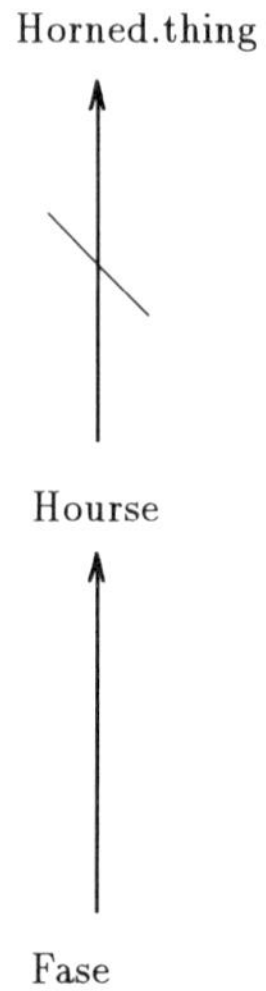

Figure 5.14: Fase, the horse

$$Fase \rightarrow horse,$$
$$horse \not\rightarrow horned.$$

To identify Fase's default property, ¬ *horned*, we need to represent explicitly that *Fase* is not an exceptional class and that ¬ *horned* is not an acquired property, using *EIR2*:

$$Horse(Fase) \odot [\neg\zeta(Fase) \wedge \neg\beta(Fase)] \rightarrow \neg horned(Fase).$$

If we add (see Figure 5.15):

$$\text{Fase} \rightarrow \text{Unicorn},$$
$$\text{Unicorn} \rightarrow \text{horse}^2,$$
$$\text{Unicorn} \rightarrow \text{horned},$$

we first identify the contradiction using *PR2*, *PR3*, and *EIR1a* (the final set of assertions is represented in Figure 5.16):

[2]Unicorns may inherit some properties of horse as well as other properties from different classes, just as griffin may inherit some properties of lion as well as eagle.

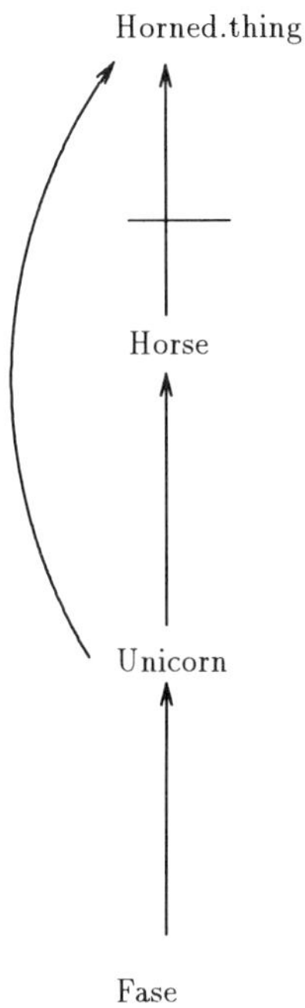

Figure 5.15: Fase, the unicorn problem

$$Unicorn(Fase) \rightarrow horse(Fase)$$

$$horse(Fase) \odot \neg Unicorn(Fase) \rightarrow \neg horned(Fase).$$

$$\odot Unicorn(Fase) \rightarrow horned(Fase).$$

$$horse(Fase) \odot Unicorn(Fase) \rightarrow \otimes horned(Fase).$$

To solve this contradiction we locate Ψ:

$$\text{Unicorn} \rightarrow \Psi,$$
$$\Psi \rightarrow \text{horned},$$
$$\Psi \bowtie \text{horse}.$$

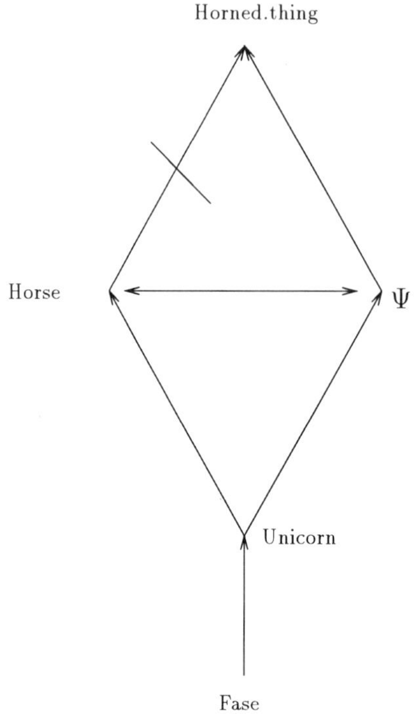

Figure 5.16: The Fase structure represented using EIR.

In order to associate the property of being horned with Fase, we need to find another path from *Fase* to ¬ horned:

$$Fase \rightarrow Unicorn \rightarrow \Psi \rightarrow horned$$

Using *EIR2* we can represent Fase's property as:

$$Horse(Fase) \odot [Unicorn(Fase) \wedge \neg\beta(Fase)] \rightarrow horned(Fase).$$

5.8 The George Problem

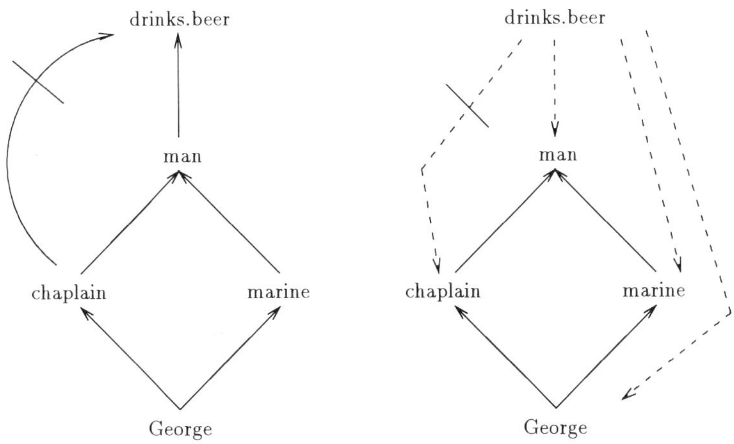

(a) The 'George' problem structure (b) EIR representation

Figure 5.17: George problem.

In the George example (see Figure 5.17a) there is a problem as to whether George does or does not drink beer. This is a real problem for nonmonotonic inheritance for which there is currently no accepted solution ((Sandewall, 1986); (Touretzky et al., 1987)). If *drinks.beer* is an acquired property (which it seems to be), then information about beer drinking must be attached to classes and individuals explicitly. The debate as to whether George is or is not a beer drinker now boils down to where *drinks.beer* should be encapsulated. If within *man*, this is not inheritable by subclasses, and similarly with *marine*, where George cannot inherit an acquired property. The 'intuitive' conclusion that George does indeed drink beer could be explained in terms of intuition deciding that beer drinking is such a strong acquired property for marines that it may be inheritable. EIR can represent such an intuition, but EIR will not allow an acquired property to be inheritable, no matter how strong the acquired property. The EIR representation of this problem structure is given in Figure 5.17b), where the dotted lines signify the choices available as to where to encapsulate *drinks.beer* (or its negation). An acquired property, because it is not inheritable, can be duplicated in more than one object. The user must decide where to locate such properties explicitly.

5.8.1 On-Path (or Acquired Properties) Versus Off-Path Revisited

Let us now demonstrate the application of EIR to the on-path versus off-path debate ((Sandewall, 1986); (Touretzky et al., 1987)). Our argument is that the two structures are not of the same topology at a deeper level when analyzed with EIR. Consider the Royal.elephant example again. Following the same analysis as made for the Clyde-Royal.Elephant example (Figures 5.1, 5.7), we obtain Figure 5.18.

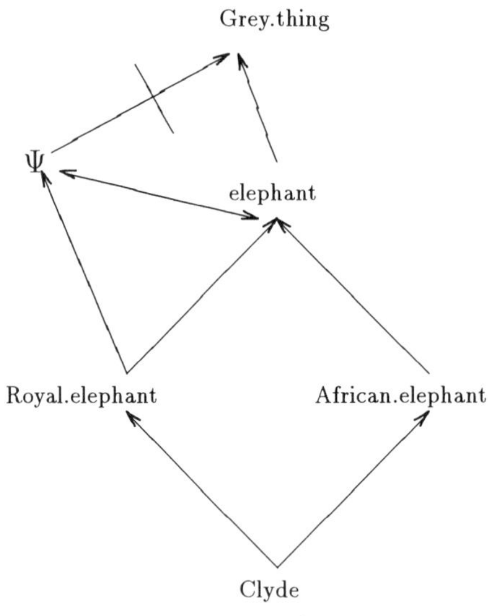

Figure 5.18: The EIR representation of the Clyde-African.Elephant structure.

Royal.elephant will inherit nongreyness through the exceptional class, just as before, as will Clyde. *African.elephant* will inherit greyness through *elephant* $\rightarrow$ *grey.thing*. The 'path' *African.elephant* $\rightarrow$ *elephant* $\leftrightarrow \Psi \not\rightarrow$ *grey.thing* is no inheritance path at all, because the bidirectional relationship between *elephant* and Ψ is a mutual (exceptional) subclass relationship and not an inheritance relationship.

Now let us look again at the George example (Figure 5.17a). In the George structure, both the classes *marine* and *chaplain* acquire their property of beer drinking from the environment. It is the explicitly acquired properties that have the last word in this structure (by EIR5 and EIR6), not the implicit inheritance properties from the class *man*. Therefore, because the acquired property of beer drinking is not associated with either *chaplain* or *marine* directly in this structure, nothing can be directly inferred about George. We are left with options as to where *drinks.beer* should be linked in as an acquired property (dashed lines in Figure 5.19).

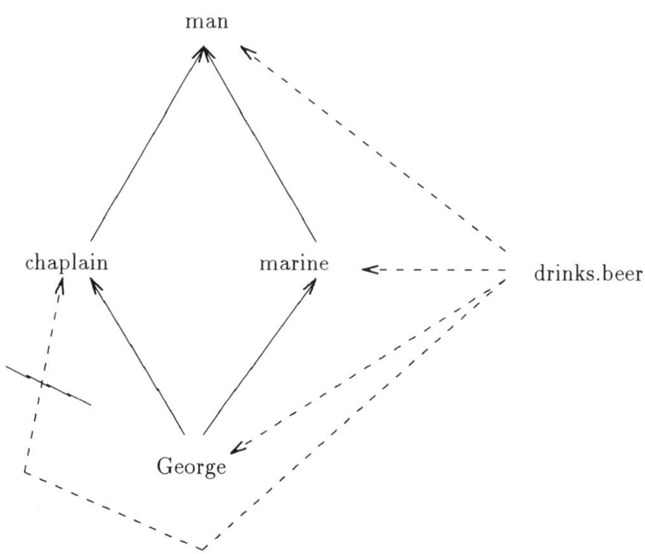

Figure 5.19: The George example represented in EIR.

The property of drinking beer can be explicitly located with *man*, *marine*, or *George*, or with more than one (because acquired properties are not inheritable). The EIR representation of the problem structure is that we have straightforward multiple inheritance, with no inheritable exceptions. If it is claimed that drinking beer is indeed an inheritable property of men (just as greyness is of elephants), this must be argued for, and that argument is a different one from on-path versus off-path debate. Importantly, the EIR representation of the *George* structure removes the implication that in certain structures more general information is to be preferred to more specific.

5.9 The Generalization of EIR

In the following sections, I adopt a generalization of EIR structures.

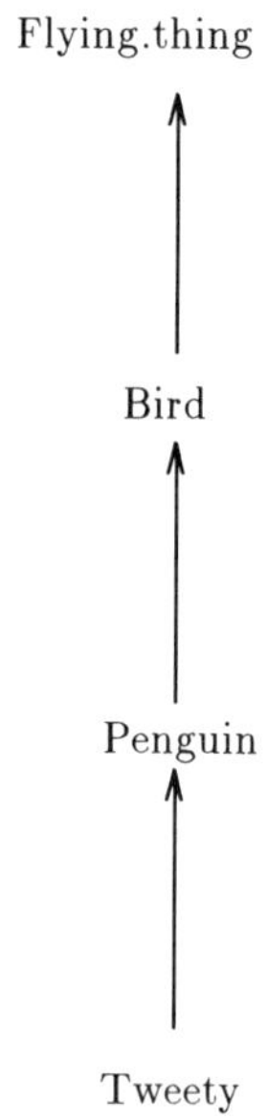

Figure 5.20: Tweety, the flying bird.

Consider Figure 5.20, Tweety is-a bird, bird is-a flying.thing, therefore, Tweety is-a flying.thing. Later, if we are told that Tweety is-a penguin, then we need to retract our previous conclusion to decide (using EIR) that Tweety, the bird, cannot fly. Although this may be considered as a basic exceptional inheritance pattern, many other structures could be more complex than that.

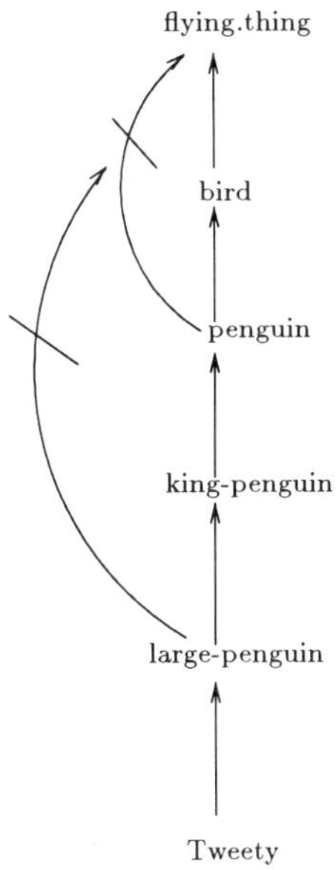

Figure 5.21: Re-exception to exceptional inheritance structure.

Consider Figure 5.21, where Tweety is-a large.penguin, is-a king.penguin, is-a penguin, is-a bird, is-a flying.thing.

A variety of exceptional paths appeared in this structure. Although penguin and king.penguin are exceptional subclasses to the superclass, bird, regarding the flying.thing property, large.penguin is an exceptional subclass to its related class king.penguin with regard to the same property. That is:

$$Bird \rightarrow flying.thing$$

$$Penguin \not\rightarrow flying.thing$$

$$king.penguin \rightarrow penguin \not\rightarrow flying.thing$$

$$large.penguin \rightarrow king.penguin \rightarrow penguin \rightarrow bird \rightarrow flying.thing.$$

This is a typical nonmonotonic inheritance hierarchy.

In order to reason with such structures, a generalization on exceptional inheritance reasoning is needed here.

An inheritance generalization could be applied as follows: Consider Figure 5.22.

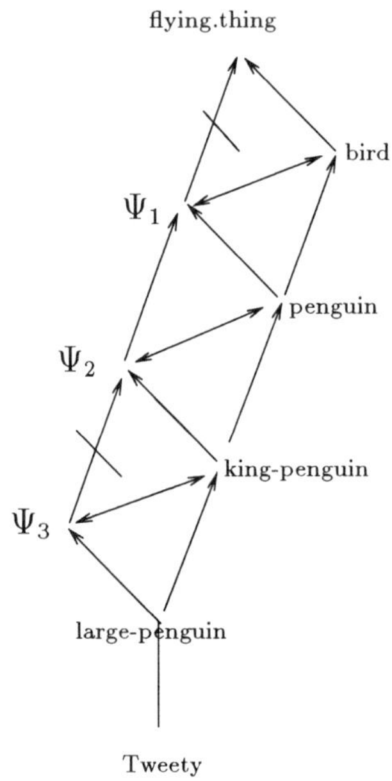

Figure 5.22: Exceptional inheritance generalization

Our starting point, at the top of the hierarchy, is to decide whether penguin is-a flying.thing or non flying.thing. Using PR2, PR3, and EIR1a (as explained before in Section 5.5 and in the examples section), penguin inherits the nonflying property from the exceptional class Ψ_1.

From the penguin node downward, we need to reason with regard to its related class king.penguin. Because king.penguin is related to the class penguin and neither an exceptional class nor an acquired property is provided to the class penguin, king.penguin inherits the nonflying property from the class penguin (and where the exceptional class Ψ_2 is compatible with its related class, Ψ_1; i.e. Ψ_2 has a positive relation with Ψ_1). Thus king.penguin not-isa flying.thing. Finally, with regard to the class large.penguin, the following observations are provided. Large.penguin is related positively to its related class king.penguin provided that an exceptional link is provided explicitly to the link penguin $\not\rightarrow$ flying.thing. That is,

$$Large.penguin \not\rightarrow (penguin \not\rightarrow flying.thing).$$

Large.penguin is an exceptional class to its related class king.penguin. Thus, Ψ_3 (the exceptional superclass of large.penguin) is an exception to the exceptional class Ψ_2, which, consequently, is identical with Ψ_1 property. Therefore, Tweety, the large.penguin, is-a flying.thing.

These exceptional classes give us a chance to state if there is an exception to the default inheritance and to the interaction between inheritance defaults. If this is the case, we can use Ψ to explain why in such situations people would like to give up the previous conclusion and change their minds.

5.10 Related Works

For the purpose of formalizing inheritance hierarchies with exceptions Reiter and Etherington (Reiter, 1980), and Etherington (Etherington, 1987b; Etherington, 1987a; Etherington, 1987c) focus on default logic. Reiter's default logic interprets the statement 'A's are typically B's' as 'if x is an A and it is consistent to assume that x is a B, then x is a B'. Exceptions create problems with typicality that cannot be handled using (normal) default logic.

Reiter and Criscuolo (Reiter, 1981) tried to solve inheritance exception by semi normal default logic. However semi normal default logic adopts the facts and the rules but not the inference mechanism behind commonsense reasoning because it needs to represent exceptional classes explicitly. In addition, semi normal default logic failed to solve ambiguity because it follows the credulous approach where no specific extension can be declare to follow.

McCarthy's circumscription (McCarthy, 1977; McCarthy, 1980; McCarthy, 1986) is based on the notion of truth in all minimal models of first-order theory. That is to say, we do not have to state explicitly the abnormality or the exceptions in one domain in order to minimize domain extension. Circumscription allows us to ignore implicit exception by assuming that it is false unless something is explicitly abnormal.

Because circumscription does not distinguish between inheritable and acquired properties, it considers all the exceptions as abnormal classes, and second because there is no distinction between explicit properties and more specific properties to allow one class to override the other, circumscription fails to formalize the appropriate commonsense reasoning that deals with exceptions effectively.

5.11 Conclusion

The chapter has provided a basic foundation of EIR. In addition I represent and identify the general role and nature of Ψ in nonmonotonic reasoning, as well as provide a precise semantics for it. The application of EIR to other inheritance problems is also being examined. Also, it is quite possible that its main application may well be in areas such as active object-oriented databases (where production rule definitions interfaced to an object-oriented database can contain exceptions where an active database is not seen as a slave to an application but as a peer that reacts to changes in the database state. In this sense, while an efficient monitoring of the active events and rules is essential, an explanation resolving conflict cases that may contain redundant information is needed), as well as object-oriented expert systems (where diagnostic reasoning needs to take exceptions into account), and analogical reasoning (where mappings must be found in object-centered representations of base and target domains). The main contribution of EIR, however, is to offer a way of restructuring nonmonotonic inheritance structures so that conflict problems can be resolved that have otherwise proved intractable. Such restructuring is based on the intuitively obvious, but often forgotten, distinctions between, on the one hand, instantiation and inheritance logic (where the two have been confused by the use of *is-a* and *not-isa* to convey both styles of reasoning), and inheritable and noninheritable (i.e., acquired) properties, on the other.

Chapter 6

Default Correlation: An Approach to Inheritance With Conflict

6.1 Introduction

Whereas Chapter 5 deals with exceptional inheritance reasoning, a procedure that is applicable to handling exceptions when *later exceptional information* is provided to an inheritance structure, this chapter deals with the other inheritance problem, namely, *ambiguity*.

Figure 6.1(a) represents the inheritance structure type that a preemption strategy could be applied to. In Figure 6.1(a) a relation (a kind of) does exist between B and C (via D) to be used by A to decide the property P, and Figure 6.1(b) represents the structure type that could be used to represent ambiguous situations.

In Figure 6.1(b) there is no (direct or indirect) relation between the classes Y and Z regarding the property P. The only relation that exists between these classes is the opposite (conflict) relation regarding one property. The aim of this chapter is to use this conflict relation in deciding a property or its negation regarding an individual, such as X. It concentrates on finding the degree of plausibility when an inheritance structure is forced to handle conflict properties related to one individual. The proposed reasoner should be able to decide which of two ambiguous situations should be preferred. Ambiguity may be defined as a situation

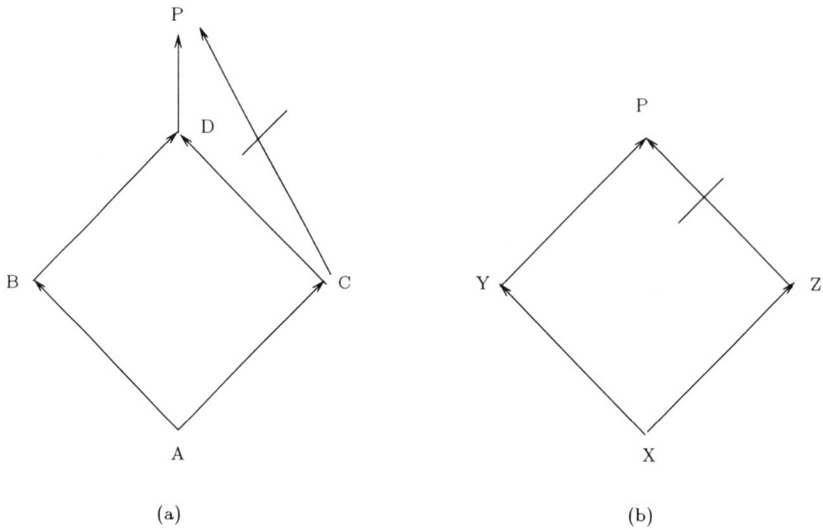

Figure 6.1: Differences between pre-emption and ambiguity

that contains several different properties that do not fit together and could be understood in more than one way. On the other hand, conflict occurs if two accounts are very different and it seems impossible for them to exist together.

We describe Default Correlation (DC), an approach to inheritance-based reasoning in knowledge representation systems for multiple inheritance with conflicts. This approach leads to a single extension when it applies to an inheritance structure. It is capable of solving the problem of ambiguity within such structures. We show how this approach deals with these problems and provide a comparison with other proposed solutions (Al-Asady, 1993b).

6.2 Inheritance

Inheritance systems focus on the procedure where properties of more specific classes are inherited from the more general ones (Brachman, 1983).

The notion of hierarchies allows instances to be organized into classes so each instance of the same class has the same properties and more specific classes inherit properties from their superclasses (Patel-Schneider, 1991). Although such systems embody an inherent nonmonotonicity (through the ability to override properties), this is not general enough to reflect real-life situations. What is needed is a more flexible inheritance theory that deals properly with nonmonotonicity (Sandewall, 1986).

6.2.1 Ambiguity

The essence of an ambiguity problem is that in multiple inheritance, when an individual or a class is related to more than one superclass, providing that each of these superclasses has an opposite property regarding the other, conflicts may occur in deciding the right inherited property by that individual. Whereas pre-emption deals with choosing a property or its negation when *later exceptional information* is provided to a related class in a structure, ambiguity deals with choosing a property or its negation from two unrelated (conflict) classes that handle these properties. Within ambiguous structures, the missing relation between typical classes (such as the relation between royal.elephant and elephant) prevents us from deciding the right extension to follow. Consider the structure in Figure 6.2.

- Republicans are not pacifist.
- Quakers are pacifist.
- Nixon is-a Republican.
- Nixon is-a Quaker.

Although Nixon, in this structure, inherits the nonpacifist property via the class Republican, Nixon inherits the pacifist property via the other conflict class (unrelated to Republican), Quaker. In the literature, there is no basis for deciding if Nixon is-a Pacifist or Nixon not-isa Pacifist because no relation between the two classes exists. In other words, in the case of ambiguous structures there is no single preferred extension, because even in the case of single-extension credulity, there is

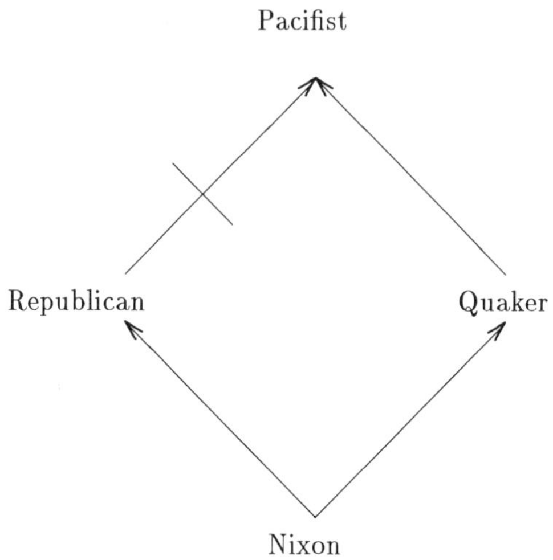

Figure 6.2: Ambiguity problem.

usually no basis for preference: the reasoning system just takes the first one that comes along [1]. (As an example, see NETL (Fahlman, 1979)).

Regardless of whether it is a path-based approach or logic-based approach, this means with credulous reasoning that we have two incompatible extensions for such structure. However, although Touretzky (Touretzky, 1986) was able to identify ambiguities (e.g., using the TINA system), he (Touretzky et al., 1987) did not suggest any solution to the ambiguity problem. Similarly, Reiter and Criscuolo (Reiter, 1981) identified ambiguity with no preferred conclusion.

Although we may consider a specific property as one of the main features to define typicality of a specific class, it is not the case to consider that all members of this typical class must have owned this property. That is, there is no way of inferring this using, for example, Touretzky's mechanism.

[1]Personal communication with D. Etherington.

To summarize this section: IDO and default logic are not capable of reasoning correctly in the presence of ambiguity.

6.2.2 Related Works

Previous work in inheritance mechanisms is explained in detail in Chapter 3, I examine some mechanisms that are related to this section. Rather than using default logic or inferential distance ordering, Neufeld (Neufeld, 1991), Bacchus (Bacchus, 1989), and Neufeld and Poole (Neufeld and Poole, 1990) adopted another approach to commonsense reasoning, namely probabilistic reasoning. Probabilistic reasoning concerns problems in which there is randomness in the world. Because with some cases it is impossible to estimate or predict the world with certainty, some statistical knowledge about the likelihood of various outcomes will help in explaining such situations. Shastri's (Shastri, 1989) evidential reasoning involves finding the most likely hypotheses among sets of alternative hypotheses. Because AI applications require reasoners that have the ability to reason with uncertain information, and because probability deals with uncertainty, probabilistic logic is a plausible inference formalization (Neufeld, 1991); (Shastri, 1989).

Whereas Reiter (Reiter, 1980) proposed a logic for default reasoning based on default rules, Reiter and Criscuolo (Reiter, 1981) described a variety of settings in which they discussed interacting normal defaults (conflicting default assumption). They claimed: "If a pair of defaults have contradictory consequents C but whose prerequests (A and B) may share common instances (a), then the typical A which is also B leads to no conclusion"(P. 97).

Brewka (Brewka, 1989; Brewka, 1991) introduced a modification of default logic called *preferred default logic*, where the notion of preferred maximal consistent subsets of the premises is used to identify preferences between conclusions. His approach subsumes Poole's approach (Poole, 1988) based on hypothetical reasoning.

All these approaches that deal with probabilistic reasoning or default reasoning in an explicit way neither allow a system to represent priorities between classes in an ambiguous situation qualitatively nor have the ability to formalize any relation between any pairs of defaults that have contradictory consequents.

6.3 Ambiguity Revisited

In our proposed approach, we start by emphasizing the nature of ambiguity within inheritance structures. Ambiguity is created when an instance or individual is related through different paths to conflict classes. Conflict classes mean typical classes with opposite properties.

Our minimal conflict structure will contain individual and typical classes with opposite properties. An individual may inherit the two opposite properties via the two conflict relations (is-a and not-isa) at the same time but with no single extension. This will create a problem in deciding the right property between the two conflict properties of such an individual.

Default correlation (DC) has a different view. It starts with the way classes are named. We could not call a class a class unless it had specific properties or components, without which the typicality of a class would mean nothing and no comparison with other classes could be made. For example, we call a human a mammal, (i.e., human is-a mammal) because some human properties are inherited through mammal properties even with some exceptions. We also call animals mammals, with some exceptions. But does that means humans are animals or animals are humans? Certainly not, because we can refer explicitly to the very obvious main differences between these classes: Humans have two legs, whereas mammals have four legs (with some exceptions). There will also be other differences, some more important than others. The importance of such differences will be called *default correlation.*

If no (direct) relation between two conflict classes exists, and in order to establish common ground (some kind of relation) between two conflict classes we need to emphasize and highlight the *indirect relation* between these conflict classes via some common property by choosing its negative and positive relation with those unrelated classes. This will be done by defining the importance (weight) of the ambiguous (conflict) property (or the level of importance of that property among other properties) regarding its related typical classes. The importance (weight) of the positive property regarding its typical class will be measured, and the importance of the negation (negative side) of the same property regarding the other conflict class will also be measured. After defining the position (level) of a specific property among other properties regarding one class, the level of the negation of the same property regarding the other conflict class will be measured as well. A comparison between the two weights (the weight of the property regarding its typical class and the weight of the negation of the same property regarding the other conflict typical class) will be measured. The comparison will be done qualitatively depending on the position of the property and its negation regarding the two conflict classes simultaneously.

In order to illustrate the concept of *default correlation levels*, we demonstrate the following example, (see Figure 6.3). In this example, we need to decide the following: Does John, who is related to two unrelated classes (namely Father class and Student class), *is-a* have.a.child or *not-isa* have.a.child?

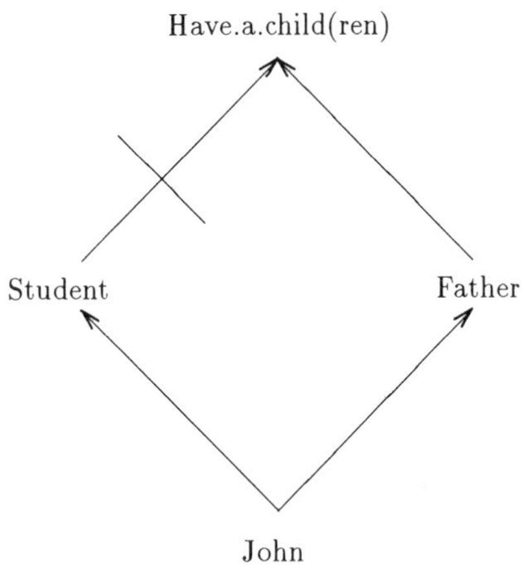

Figure 6.3: Two conflict classes regarding one property.

According to DC, we analyze the two conflict classes to their properties. For example, the typical class Father could be analyzed to the following properties (features)[2]:

1. Have (at least) one child.

[2]Features mean the subproperties or the components that can create a class name.

2. Adult.
3. Have a wife.
4. Have a job.

Obviously, the arrangement of these properties has been done according to their levels of importance to the Father.class; having a child is more related to the typical class Father than the property of having a job.

Similarly, we can analyze the typical Student.class to the following properties:

1. Register in an academic site.
2. Irresponsible.
3. Possibly (possibly not) have a job after study.
4. Possibly (possibly not) have a wife.
5. If have a wife, possibly (possibly not) have a child.

Thus, to decide whether John has a child or not, we compare the levels of having a child between the two classes at the same time. Whereas having a child within the student.class weighted W_5, having a child within Father.class weighted W_1. We conclude that John is-a having a child property.

In Figure 6.4 we represent 'have a child(ren)' problem in terms of a balance. We put the positive property in the first dish and the negation of the same property in the second dish as weighted factors regarding the conflict classes. The heavier side (more important to its class) will outweigh the other (the less important to its class). We could imagine the balance acting like the scales of justice, where the levels act rather like weights in a balance allowing us to select the correct extension. The difference will determine if there is a (stronger) inheritance relation between one of the conflict classes and the superclass. Contrast this with previous solutions to the ambiguity problem (e.g., Section 6.2) in which all evidence is given the same weight - there is no way of denoting important information as being more relevant than trivial when searching for a path.

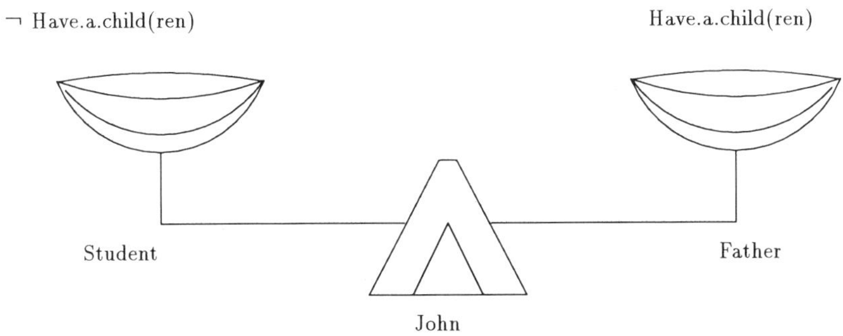

Figure 6.4: Conflict balance.

Summarizing the general notion of DC, the importance of a property to its class is used to weight the evidence for and against the negation of the same property to the other conflict class. One side will outweigh the other in this network and that will decide the path we must choose.

Before proceeding to describe the default correlation framework and define the procedure that we use to solve ambiguity in detail, there are two issues that need to be discussed and clarified here;

1.How do humans as intelligent entities, use their commonsense reasoning to deal with conflict situations?

2.What is the measure that humans use in evaluating conflict classes with each other and the relation between two conflict classes in preferring one class or the other?

In an inheritance structure, the core problem of ambiguity is that ambiguous classes regarding a property have no relation with each other but have separate relations with their unique related individual. For example, if I say 'a specific text is readable by me, and the same text is unreadable by me', that means readability and unreadability are gathered together within one superclass, namely *the specific text.* That means there is somehow an *indirect relation* between the two conflict classes (readable text, unreadable text) because they are gathered together within one joint class, *text.*

In order to measure the indirect relation, we need to analyze each of the conflict classes separately regarding its superclass (i.e., to assign the class's properties regarding the direct superclass, such as text in the previous example).

In real life, humans as intelligent entities perform such comparisons by understanding the nature of each of the conflict classes separately, and jointly. That is, what are the properties of a class alone and what is the effect from other conflict classes on these properties? All such comparisons and eventual property preferences will be done by default. The successful default (preferred) among others will rise (according to its priority) to deal with such comparison.

For instance, if we lose somebody in a jungle, we prefer to find that person first rather than take our meal even if we are hungry (notice there is no obvious relation between the hungry.person class and lost.somebody class). Or, we prefer a glass of water in a desert rather than a rest for 2 hours under a tree (although there is no obvious relation between water and resting). Also, we may prefer to consider a person who seems guilty of committing a crime innocent if we take into account their excellent record so that we can reason that the event was accidental.

Consider Figure 6.5. The aim of this figure is to illustrate the different kinds of ambiguity. Because ambiguity can be approached from different levels, our approach will ignore the ambiguity at the definition level because there is no way at such a level to decide the right extension. Also we ignore the level in which we reach ambiguity by default, such as recognizing a property in one class and its negation implicitly (not explicitly) by the other. Our aim is to reach ambiguity and try to represent it using two conflict classes regarding one explicit property.

In Figure 6.5(a), obviously the conflict property is have.a.child in which John either will be a member of the have.a.child class or not. In Figure 6.5(b) there is a different question: What is the conflict property between lost.son class and hungry class? It seems that as we create the top-level property that has a specific (e.g., positive) relation with one of the conflict classes, an opposite (e.g., negative) relation with the other class will be created. That means, with regard to our example, as soon as we identify a specific relation that relates either hungry class or lost.son class with its top level property, an opposite relation between the other conflict class and the top level property will be assigned directly. Therefore, in order to assign a property to the structure in Figure 6.5(b),

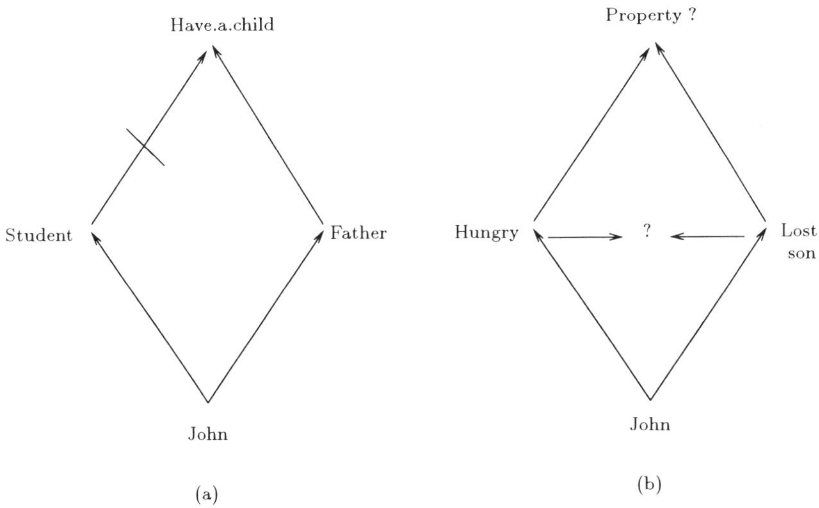

Figure 6.5: Different kinds of conflict structures

an analysis of the features of the conflict classes is needed. The top level property (in different level and different sign relation) will appear within such analysis.

Because it is hard to create numbers or percentages to assign the importance of a property to a class[3] we use a form of qualitative reasoning based on quantitative measures. What is meant by 'qualitative' here is the degree of importance of a specific property to one class as compared with another class' properties.

6.3.1 Default Correlation Framework

In this section I describe an approach to nonmonotonic inheritance that is capable of handling ambiguity problems. The intuition underlying this approach is that humans, as intelligent entities, deal with the conflict properties by employing a kind of default reasoning. This is Default Correlation (DC).

[3]The importance of a property to a class will differ from domain to domain.

Definition 18 *Default Correlation (DC) Is the method of resolving a conflict situation by analyzing each of the conflict classes in terms of properties or weights (where the conflict property that creates such conflict situation must be included). More important properties regarding one class (positively or negatively) will be located at higher levels of the analyzed class, and less important properties will be located at lower levels.*

Within one class, each property among others represents a default correlation level (weight) that assigns its importance, through its order, to a class. DC levels regarding one class are subject to change from one structure to another according to the other conflict class and the conflict property.

Corollary 1 *Because the importance of each level regarding its related class has a different order in different structures (regarding other conflict classes), DC relations are defeasible.*

In order to analyze one typical class for its levels or properties, assuming that each class is divided into n properties, the following procedure is used:

Analyse the two conflict classes, regarding one conflict property, for their sub-properties or features. Divide these features according to the degree of importance to a class to *more distinctive properties* and *less distinctive properties.* Order these features according to their importance, including the specific conflict property in each, in a specific order (either descending or ascending). Compare the level of the conflicting property between the two classes. The higher level regarding the conflict property in one class will override the lower level of the same property (or its negation) in the other.

The first half of the ordered properties ($n/2$) in each class will be called *properties with high correlation weights.* They represent features that play an effective role (important role) in defining and distinguishing a class from other classes.

The other half of these ordered properties, ($n/2$), will be called *properties with partial correlation weights.* They represent properties that have less influence (are less important) when comparing a class with other classes.

Rule 1 *Default Correlation Rule: High correlation level properties override lower (partial) correlation level properties when choosing a path among others through a network.*

$DC[weight_i(class_1), property_0] > DC[weight_j(class_2), \neg property_0]$ *where* $i > j$.

That is, the default correlation between $weight_i$ of $class_1$ and $property_0$ is higher than the default correlation between $weight_j$ of the class $class_2$ and the negation of the same property $property_0$. In other words, the order of $weight_j$ is below the order of $weight_i$ in the default correlation table.

For example, if an instance has two relations with these two conflict classes, then the structure extension will go through $class_1$.

We now give an informal description of what we mean by properties with high correlation weights and partial correlation weights.

Definition 19 *Default correlation weight*

It is the qualitative level ($\mathcal{W}_i$) *for the relation between a class and one of its properties. The different levels of properties are represented as follows:*

1. $\oslash$ *weight represents a positive but weak relation between a class and one of its related properties. We represent a link with such relation by* $\xrightarrow{\oslash}$.
2. $\neg\oslash$ *weight represents a weak and negative relation between a class and one of its related properties. It is similar to* $\oslash$ *except it represents a negative relation. We represent a link carrying such relation as* $\overset{\oslash}{\not\longrightarrow}$.
3. $\oplus$ *weight represents a positive and strong relation between a class and one of its related properties. We represent a link with such weight as* $\xrightarrow{\oplus}$.
4. $\neg\oplus$ *weight represents a highly negative relation between a class and one of its related properties. It is similar to* $\oplus$ *except it represents a negative relation. We represent a link carrying such weight as* $\overset{\oplus}{\not\longrightarrow}$.

A highly related link can override a partially related link in a conflict situation; that is, $\oslash$ and $\neg\oslash$ links are subject to be overriding by $\bigoplus$ and $\neg\bigoplus$ links.

In multiple inheritance with conflict, we use different levels of a theory to represent different degrees of reliability.

A DC theory T is a set$(T_1 ... T_n)$ where each T_i is a set of weights $(W_1 ... W_n)$ inside a class. The information in $\bigoplus W_1 T_i$ is more reliable (distinctive) than $\oslash W_n T_j$.

There is an ordered set of weight (W_i) levels attached to the links. These weights are ordered according to their relation with the class. Highly correlated links or highly excepted links are attached to the highest weight level in the network, whereas partially correlated links or partially excepted links are attached to the lowest weight level in the network ($\bigoplus W_1$ or $\oslash W_n$). Links are compared with each other (for the same class or different classes) according to the following procedure:

$W_1 > W_2 > ... > W_n$.

Similarly

$\neg W_1 > \neg W_2 > ... > \neg W_n$.

$\bigoplus$ or $\neg\bigoplus$ overrides both $\oslash$ and $\neg\oslash$ relations.

6.3.2 Ambiguity Revisited with DC

In order to illustrate the concept of DC levels, I use a modified version of the George example ((Touretzky et al., 1987); (see Figure 6.6)). George is an individual with two links: George *is-a* marine as well as George *is-a* chaplain. The class marine is related positively to the property *getting-drunk*, but the class chaplain is related negatively to the same property. This leaves George in a conflict situation. We now show how DC deals with the ambiguity problem. Consider Figure 6.4 again. To decide if George is a 'getting-drunk' person or not, we analyze the classes chaplain and marine, regarding the getting-drunk property, to their DC weights. These weights (levels) are given in Table 6.1. The features of each class, including the conflict property, are listed in a *descending order* according to their importance. In order to find out whether George is getting-drunk or not we compare the level of getting-drunk property in the first class (marine) with the level of the same property in the other class (chaplain).

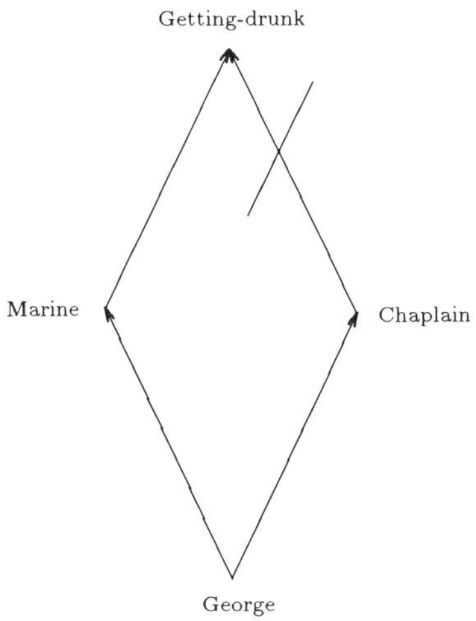

Figure 6.6: partial and high links

This is our view in explaining the way intelligent entities can deal with conflict situations. Because W_{c1} is one of the highest level features to the class chaplain, it will be represented by $\oplus$. Similarly, and because $\neg W_{m4}$ is one of the features that is less important in defining the class marine, it will be translated into $\neg\oslash$. In Table 6.1, W_{c1} has a higher correlation level than $\neg W_{m4}$. Such DC levels will be translated into high or partial qualitative default, as shown in Figure 6.7.

An argument against this way of ordering features may arise but we could argue by asking the following question: To call a marine a typical

Chaplain	level	Marine	level
1 Does not get-drunk	W_{c1}	1 Obeys orders	$\neg W_{m1}$
2 Pray regularly	W_{c2}	2 Able to fight	$\neg W_{m2}$
3 Honest	W_{c3}	3 Wears uniform	$\neg W_{m3}$
4 Telling truth	W_{c4}	4 Getting-drunk	$\neg W_{m4}$

Table 6.1: Default Correlation Levels.

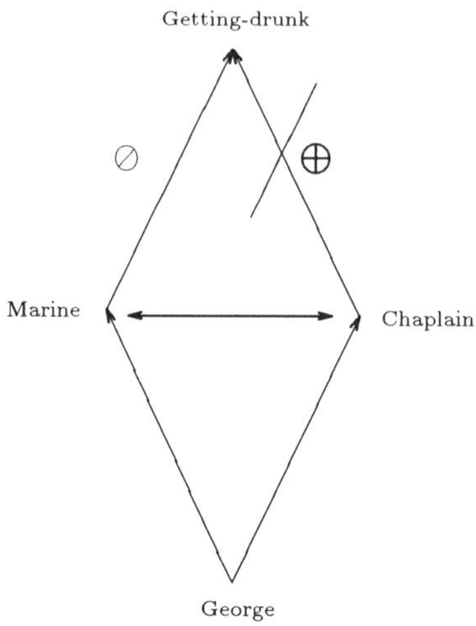

Figure 6.7: Default correlation's links.

marine, which is more important: for the typicality of this class to have an obey orders feature (by default) or to have a getting-drunk feature? Obviously, the obey orders property can define the class marine in a better way than the getting drunk property. The obey orders property is a more distinctive property to the class marine than the property of getting drunk. Thus we may attach a weight to each property, among others, when there is a need to define a priority measure between them regarding one class.

This priority between defaults has the same influence on the typicality of any other class. For example the relation between the class 'driver' and the property 'know how to drive' has a higher relation than some other features to the same class. Also the relation between the property 'have children' and the class 'Father' is higher than the relation 'have a job' and the same class 'Father'. I.e., the property 'have children' can define the class 'Father' in better way than the property 'have a job'.

In general, it is the priority between different (default) features regarding the typicality of the same class, that is used to disambiguate any conflicts that arise in multiple inheritance due to one class being a specialisation of another.

6.4 Default Correlation Framework Algorithm

In this section, an algorithm to demonstrate the mechanism behind default correlation is represented (see Figure 6.8). This algorithm started by identifying if there is a relation between the conflict classes. If there is such a relation, then the EIR algorithm of Chapter 5 is applied. Otherwise analyze each of the conflict classes for its different weights and properties. Arrange these properties according to their importance in each class. Compare the level of the conflict property between the two conflict classes. Choose the highest level, among other levels, as the right extension to follow. The right level indicates the right class, among others, to choose.

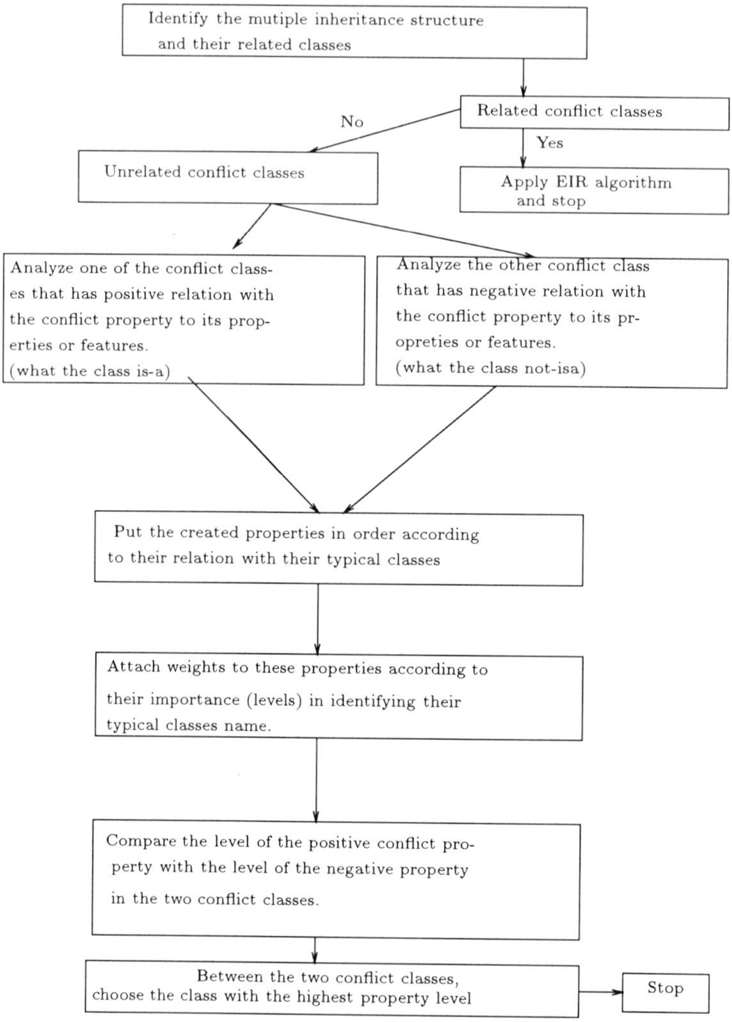

Figure 6.8: DC Algorithm.

6.5 A Formal Description of the Representation Language

In this section we represent a new language in order to formalize the DC system.

We have four kinds of relations and four kinds of entities:

$\ell = \langle \mathcal{C}, \beta, \mathcal{W}, \delta, \Lambda, \ll, \Omega, \Sigma \rangle$.

where

- $\mathcal{C}$ is an instance.
- β_i, $i=1,\ldots,n$ represents a class in a conflict situation.
- $\mathcal{W}_{\beta_i}$, $i=1,\ldots,n$ represents the weight of the property (i) in the class β.
- δ represents a conflict properties set.
- Λ is the subset of $\mathcal{C}$ that consists of all possible extensions of δ, that is, mapping from $\mathcal{C} \times \delta$ to the extension.
- $\ll$ is the partial order relation between the weighted factors within one class.
- $\Omega(\mathcal{W}_i)_{\beta_i}$ is the relation between a class β_i and its weighted factors, or all possible ordered levels of β regarding δ, that is, the weighted factors in β_i.
- $\Sigma(\delta)_{\beta_i}$ is the level of the conflict property δ regarding the class β_i.

We could demonstrate this language with the following representation example.

6.5.1 Example

We could represent George, the marine-Chaplain problem, as follows:

Let:

- $\mathcal{C} \rightarrow$ George.
- $\beta_i \rightarrow$ Chaplain.

- $\beta_j \rightarrow$ Marine.
- $\delta \rightarrow$ (Getting-drunk, $\neg$ Getting-drunk).

Then:

$\Lambda = (\mathcal{C} \times \delta) \rightarrow \{Chaplain \not\rightarrow Getting.drunk, marine \rightarrow Getting.drunk\}$,
$\Omega(\mathcal{W}_i)_{\beta_i} \rightarrow (W_1 \ll W_2 \ll ... \ll W_n)_{\beta_i}$
$\Omega(\mathcal{W}_i)_{Chaplain} \rightarrow$ (does not getting-drunk $\ll$ pray-regularly $\ll ... \ll$ $telling - truth)_{Chaplain}$
$\Omega(\mathcal{W}_i)_{marine} \rightarrow (obeys - orders \ll abletofight \ll ... \ll$ $getting - drunk)_{Marine}$
$\Sigma(\delta)_{\beta_i} \rightarrow \Sigma(Getting.drunk)_{Marine} = 4$
$\Sigma(\delta)_{\beta_j} \rightarrow \Sigma(\neg Getting.drunk)_{Chaplain} = 1.$

Because the weight of the getting-drunk property regarding the class marine is lower than the weight of the $\neg$ getting-drunk property regarding the class Chaplain, we will choose the extension: $George \rightarrow Chaplain \not\rightarrow getting.drunk$.

6.6 Conclusion

In this chapter a new approach to nonmonotonic inheritance with conflict is presented. We dealt with the levels of conflict classes in qualitative respect and without recourse to the quantitative manners. A new language within this approach is also presented.

Chapter 7

Application: Causal Reasoning and EIR

7.1 Introduction

The aim of this chapter is to highlight the role of inheritance structures in defining causal relations. This is done by identifying the relation between the top-level properties of two inheritance structures. Consider the following three levels of an inheritance hierarchy:

(a) Individuals,

(b) classes and subclasses, and

(c) properties.

In such a hierarchy, individuals at the bottom level share the top-level properties via their related classes and subclasses in the middle of this hierarchy. The following path represents such a relation:

$$my.glass \rightarrow glass \rightarrow fragile.thing,$$

where *my.glass* is an individual that inherits the property *fragile.thing* via the class *glass*. In this path, although the class *glass* is-a *fragile.thing*, *fragile.thing* is not a property for just this class. Instead it is an abstract for a set of classes sharing the same property. For example, in addition to the class *glass* which inherits the *fragile.thing* property we can refer to a set of other classes (such as watches, bulbs, TV, computer terminals, and many other similar classes) that inherits the same property. That is:

$$[glasses, watches, bulbs, ...] \subset fragile.thing.$$

These classes are a subset of the fragile.thing set, (see Figure 7.1).

Similarly, this analysis may be applied to the following inheritance path:

$$my.floor \rightarrow floor \rightarrow hard.thing$$

From these inheritance paths we may create a causal relation. Combining top-level properties of two inheritance structures by an *action* will introduce and create a causal relation. For example, throwing *my.glass* on *my.floor* causes *the breaking* of *my.glass*. In this example, we can recognize the inheritance structures and the causal relation between them. Namely, the structure that considers *my.glass* as one of its instances; the structure that considers *my.floor* as one of its instances, and the causal relation between them. The relation between different inheritance structure properties can play an important part in defining causal structure and causal reasoning.

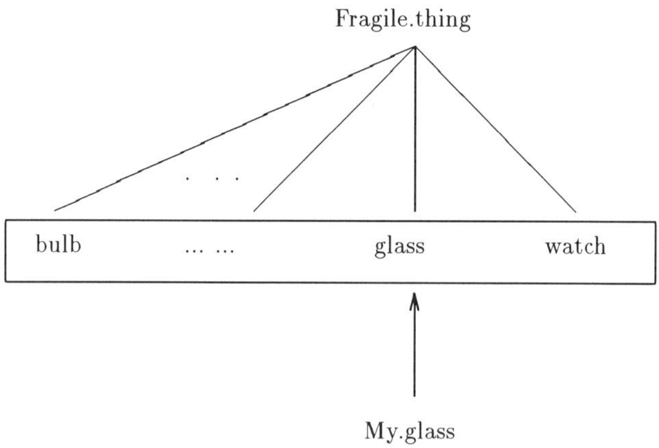

Figure 7.1: A property with multiple classes

My goal in this chapter is to define causal relation as a relation between top-level properties of two inheritance structures rather than between classes or instances of the same structures.

7.2 Causal Reasoning: Artificial Intelligence Issues

Modern interpretation of causation can be traced back to the work of Lewis (Lewis, 1973), who defined causation according to counterfactuals theory. A counterfactual statement is a conditional statement whose antecedent is false, or thought to be false. It is a statement of the form *if ... then.* Lewis defined counterfactual dependence as a relation between two families of possible events (or propositions in general), as in Event X caused Event Y, or a relation between event types, as in events of Type X cause events of Type Y. Although Lewis (Lewis, 1973) rejected the temporal direction in causal relation, he wrote:

> 'I reject that a cause must always precede its effect ... since it rejects a priori certain legitimate physical hypotheses that posit backwards or simultaneous causation'.

He considered events here to be entities located (implicitly) in time in the way that 'if X and Y are events, and X is the cause of Y, then there is a corresponding counterfactual that holds, of the form 'if X had not occurred, then Y would not have occurred.' Lewis added to the notions of possible worlds and the accessibility relation between them the notion of similarity between these worlds that could determine the 'distance' from World 1 to World 2. (I.e., Event E causally depends on Event C, just in case whether E happened or not depends on whether C happened or not).

Shoham (Shoham, 1988) reversed the idea of Lewis; he suggested reducing the notion of counterfactual implication to the notion of causation rather than vice versa. Shoham (Shoham, 1988) presented an analysis of causal reasoning from cause to expected effect, from knowledge and ignorance about the past to knowledge about the future. A causal theory is a collection of causal statements, which are rules of the form:

$$\Box\phi_1 \land \Box\phi_2 \land \ldots \land \Box\phi_m \land \Diamond\chi_1 \land \Diamond\chi_2 \land \ldots \land \Diamond\chi_n \rightarrow \Box\psi$$

where the modal operators $\Box$ and $\Diamond$ intend to mean *necessity* and *possibility*, respectively, ϕ_i, χ_i and ψ are atomic sentences expressing the active cause, the background conditions, and the effect, respectively. For example:
$\Box Press-button(t) \land \Diamond All-working(t) \rightarrow \Box Bell-rings(t+1)$ (Galton, 1991)[1].

The key assumptions underlying Shoham's theory are that causality is an epistemic notion and causal reasoning is inherently a kind of nonmonotonic reasoning, Galton (Galton, 1991) criticized these assumptions. Galton questioned the key assumptions underlying this theory and the complete failure of Shoham's theory to take account of the state-event distinction. (A. Galton, Personal communication.) Although questioning this theory, Galton (Galton, 1991) offered no alternative theory of his own.

Although Ginsberg (Ginsberg, 1986) admitted the close relationship between counterfactuals implication and causality, he did not share the same outlook (such as in (Lewis, 1973)) regarding the connection between them. He wrote: 'There is a close but incomplete relation between counterfactuals implication and causality. This connection cannot be pushed too far, however. In addition, there are counterfactuals whose entire purpose is to indicate that the premise and conclusion are not causally linked. In general, it does not seem possible to develop a formal connection between counterfactuals and causality.' (Ginsberg, 1986).

Haugh (Haugh, 1989) investigated the use of nonmonotonic temporal causal logics for formalizing temporal persistence and for characterizing commonsense inference patterns.

Suppes (Suppes, 1970) proposed a probabilistic approach to causation; that is, the probability of one event causing an effect. He identified different kinds of causation, such as direct cause and indirect cause, sufficient and negative causes, and supplementary causes (Shoham, 1988).

Reiter (Reiter, 1987) identified 'effect' as an observed fault represented by a first-order sentence, and 'cause' or 'diagnosis' as a minimal

[1]Galton, among others, questioned the actual meaning of $\Box$ and $\Diamond$ in Shoham's theory.

set of 'abnormal' circuit components whose abnormality is sufficient to produce the fault. Cox and Pietrzykowski (Cox, 1986) presented an analysis of causation in the context of fault diagnosis with no explicit representation of time. Causes correspond to first-order sentences of a particular form, and are relative to a context that consists of a set of first-order axioms that describe the behavioral characteristic of the system under diagnosis.

Finally, Pearl (Pearl, 1988) questioned the ability to formalize a system to reflect commonsense inferences without acknowledging the distinction between causality and default logic. Pearl claimed that any commonsense reasoning involves two types of default rules: causal rules or expectation evoking such as 'if fire then smoke' and evidential rules or explanation evoking such as 'if smoke then fire.'

The remainder of this chapter is concerned with presenting a different view of causation, one that uses the concepts of EIR.

7.3 Relation Between Inheritance Structures and Causal Structures

In this section I demonstrate the relation between inheritance structure and causal relation. In other words, the aim is to use inheritance structures to assist in analyzing and formalizing causal relations. Consider the following statements: turning the ignition key causes the starting of its attached engine, throwing a glass causes it to break, applying a flame to a faulty gas point causes an explosion or fire. These statements contain an action (or event) applied to an object.[2] Turning', 'throwing' and 'applying' are different actions applied to the 'ignition key', to a 'glass' and to a 'faulty gas point' respectively. We may consider these objects as instances of specific classes, each with a specific property, i.e., a glass is an instance of the class 'Glass' with property P, such as 'fragile', (see Figure 7.2).

Consider another example. My pencil has a broken lead, which prevents me from using it. Sharpening my pencil will enable me to use it. My intention in these examples is to show that the same pencil has been

[2]A clear distinction between the terminologies action, state, cause, and effect, among others, is provided later in this chapter.

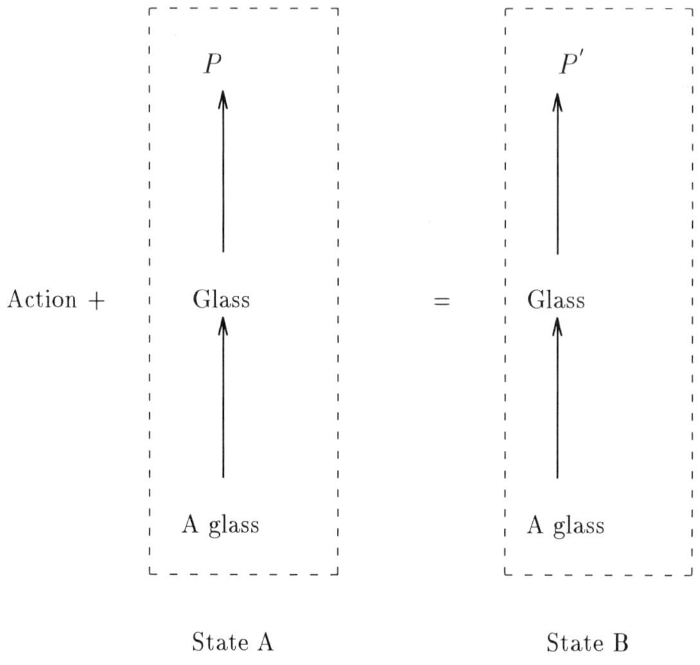

Figure 7.2: Action (or event) applied on the instance 'A glass' causes its property P to be P'.

in more than one state (property), namely un-usable and usable pencil. Similarly, filling my pen with ink allows me to use it. In other words, causation through an event (or an action) moves an individual from one state to another state, (i.e., from unusable pencil to usable pencil). Thus, the states of an individual express the different properties that an entity possesses at different times. Figure 7.3 represents the many states that an instance may possess at different times via an event.

An event (action) can change a state of a class:

Action: *throwing* a glass $\xrightarrow{Causes}$ State_1: *unbroken-glass* $\xrightarrow{To-be}$ State_2: *broken-glass.*

Figure 7.4 represents the role of such actions.

State-1	State-2	...	State-n
Property-1	Property-2	...	Property-n

Figure 7.3: The states of an individual.

Consider the following two sentences:

'Throwing 'my.glass' on 'my.floor' causes the breaking of my.glass' and 'Fire causes smoke.' In order to study and analyze the causal structure behind these two sentences, we need to split each sentence into an action and states. In the first sentence, the (explicit) action is 'throwing'. This action acts (relates) between the two instances 'my.glass' and 'my.floor'. Whereas 'unbroken-glass' represents state$_1$ of the instance 'my.glass' (by default), 'broken glass' represents state$_2$ of the same instance. Bear in mind that the action happened between two objects, namely 'my.glass' and 'my.floor', in which the same role for 'my.floor' needs to be recognized as well. In the second sentence, there is no explicit action, such as 'applying', that we need to determine by default. This implicit action acts between two instances. Whereas the first instance is defined explicitly, 'fire', the other instance is an implicit one. We may represent this instance by the symbol X. Under such representation, a default property, such as 'unsmoked-X', will represent state$_1$ and 'smoked-X' will represent state$_2$. Thus, we may clarify the default action behind the 'fire causes smoke' statement in a general sentence as follows:

'Applying fire on object-X causes object-X to produce smoke and something else.'

For instance:

- 'Applying fire on 'gas' causes the turning of the 'gas' to 'smoke' and 'flame.'

- 'Applying fire on 'wood' causes the turning of the 'wood' to 'smoke' and 'heat.'

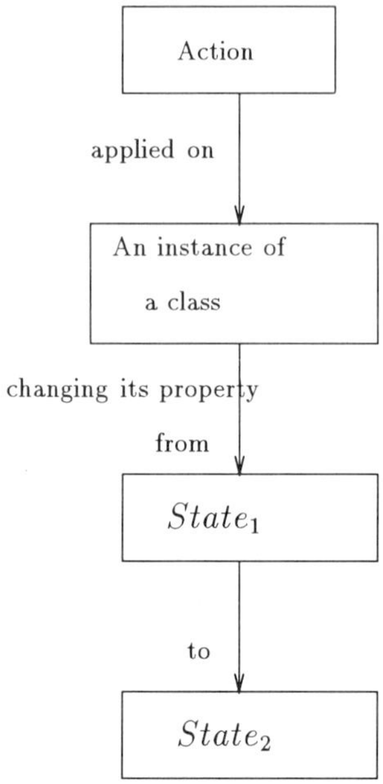

Figure 7.4: The role of an action on the properties of an individual

- 'Applying fire on a 'book' causes the turning of the 'book' to 'smoke' and 'ash.'

To summarize; in order to define a causal structure, we distinguish between two kinds of actions, namely *implicit (default) action* and *explicit action*. For example, 'fire causes smoke' consists of an implicit action that intends to mean 'applying' fire (an instance of one structure) on X (an instance of a second structure) causes smoke. On the other hand, 'throwing my.glass on my.floor causes the breaking of my.glass' includes an explicit action.

The question here is what type of structure has been used behind such causal statements. Consider modified versions of the broken glass example: 'Throwing my.glass on my.floor causes the breaking of my.glass' and 'throwing my.glass on my.floor causes the cracking of my.glass'. In these examples, we can recognize the different states of 'my.glass' , namely 'cracked' and 'broken'. Those various states demonstrate the different situations that an instance may be in via a specific action [3].

The question that arises here is because we are dealing with the same instances 'my.glass' and 'my.floor', why did we get the breaking state in the first one and the cracking state with the other? In other words, what is the basis for recognizing the different states that an object will be in if applying the same action? To answer this question we need to define the structure(s) behind a causal sentence. We may consider a causal relation as a relation between the top-level (properties) of two inheritance structures. At the bottom level of such structures, we have instances related to the top-level (properties) via a specific class. Regarding the given examples one of these structures involves the instance 'my.glass' which is related to the property 'breaking.thing' via the superclass 'glass', but the other involves the instance 'my.floor' which is related to a default property, such as 'hard.thing', by the superclass 'floor' [4].

Figure 7.5 represents such a relation, where my.glass is an instance of the class 'glass' related to the property 'fragile.thing'. Similarly, the instance 'my.floor' is an instance of the class 'floor' with the property 'hard.thing'.

7.4 Causality and EIR

7.4.1 Nonmonotonic Causation: EIR Explanation for Causal Structure

Consider the following examples:

- Breaking the battery prevents the car from moving.

---

[3]The conditions that have a chance to influence causation, such as defining the distance that we threw my.glass, or the way that we threw it, or any other circumstances when an action occurs, are discussed later. My intention here is to define the multiple states (properties) that an individual enjoys.

[4]What is implicitly used here (using EIR from Chapter 5) is the notion of *typicality*. A (typical) glass dropped on a (typical) floor from a (typical) height (assuming the consistency of the other background conditions χ in order for the active cause to have its effect) has a good chance of breaking.

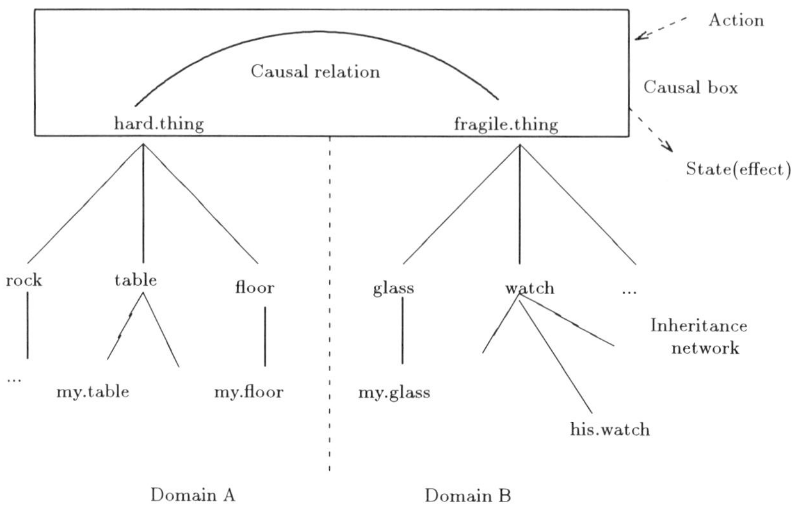

Figure 7.5: Causation via inheritance.

- The dead battery prevents the car from moving.
- Replacing the real battery with a toy battery prevents the car from moving.

As mentioned before, in order to create a machine with the ability to reason in real-life domains, it is essential to distinguish between two types of events, namely an *explicit action* (or event, happening; (Shoham, 1988)) and an *implicit action*. In the first example, it is obvious that the action 'breaking the battery' prevents the car from moving. In the second example, we introduce a state before adopting the action. In this example, there are many reasons for a battery to be dead (maybe the acid is too old or we kept the lights on all night or the battery was damaged for some reason, etc.) We can describe this state as an indication of an implicit (default) action. In other words, it is the property (state) that represents a cause behind an implicit action.

What about the third example, namely the battery is a toy battery? The cause behind preventing the car from moving is obvious, but there is no explicit action here. The car contains a toy battery or the standard battery replaced by a toy battery; all are states to a unique action. Namely, the existing battery is not a normal (a standard, a default, a typical) battery.

Accordingly, we can say the abnormality (of the battery) [5] is one of the battery states that hides an actual action behind. This could be represented as follows:

(default action) replacing real battery with a toy battery, $\text{state}_1 \rightarrow$ the battery is not a normal battery, state_2 (effect) $\rightarrow$ the car is not moving.

We can summarize the above state-action (event) relation as follows: An *event* can cause another *event* or a *state*, and a *state* can allow or prevent an *event*.

Event: Breaking battery $\overset{causes}{\longrightarrow}$ State: Battery broken $\overset{prevents}{\longrightarrow}$ event: Car starting.

Event: Leaving lights on $\overset{causes}{\longrightarrow}$ State: Battery dead $\overset{prevents}{\longrightarrow}$ event: Car starting.

Event: Replace with toy battery $\overset{causes}{\longrightarrow}$ State: No proper battery $\overset{prevents}{\longrightarrow}$ event: Car starting.

Thus, to say that an event prevents another event is to say that it causes a state that prevents that event.

I examine one of the features of the defined relation between causal relation and inheritance structures, namely the ability to reason in the *absence* of complete information.

I use EIR (explained in Chapter 5) to discuss nonmonotonic causation. In real-world situations, as we have seen with inheritance structures, there is no rule to follow without exceptions. Because a causal relation is created as a relation between states (properties) of inheritance structures, causal reasoning must inherit its reasoning mechanism from the used inheritance structures. Following the exceptional feature of inheritance structures, a causal structure is subject to exceptions as well. As defined earlier, a causal relation is a relation between an action and an expected state (effect).

[5] At this point we may think of the abnormality of an object as one of two states: inheritable state, which is a state that it is possible for an object to have in normal life; and an acquired state, which is the state that is not possible to happen in normal life unless specific circumstances have occurred, such as in the toy battery example.

Consider the following sentences:

Throwing 'my.glass' on 'my.floor' fails to cause 'my.glass' to break.

Throwing 'my.glass' on 'my.floor' causes 'my.glass' to fly.

We need to reason why we did not get the expected result, that is, 'the breaking of my.glass'. There is more than one possibility to explain why the expected state does not occur, that is, does not inherit its property 'fragile.thing' by default[6]. It is the reality (the exceptional cases) that disallows similar classes or individuals with a similar structure to follow the same rule. We explain such exceptional behavior (or state) as a consequence to structural behavior in the base, namely exceptional inheritance structures. For example, 'my.glass' is-a 'plastic.glass' or 'my.glass' is-a 'toy.glass' or 'abnormal glass'. From an inheritance reasoning point of view, and because the exceptional classes 'plastic.glass' and 'toy.glass' are considered to be subclasses of the superclass 'glass' regarding the default 'fragile.thing', causal structure on the top of inheritance structure must inherit this behavior explicitly. That means we have an exceptional relation between a subclass, such as 'toy.glass', and the default property 'fragile.thing'. Consequently, applying action between ¬'fragile.thing' and 'hard.thing' will produces ¬ 'breakable' state. Figure 7.6 represents such exceptional behavior.

$$my.glass \rightarrow toy.glass \rightarrow \Psi \not\rightarrow fragile.thing$$

or

$$my.floor \rightarrow toy.floor \rightarrow \Psi \not\rightarrow hard.thing.$$

In other words, my.glass and my.floor did not inherit their relative properties by default. Those individuals are exceptional instances related to their properties through the exceptional class Ψ.

[6]It is important to stress here the role of typicality in that properties like hard.thing and fragile.thing are *relative*. An object cannot be regarded as fragile absolutely but only in a context: In this sense it is just like 'big' (e.g., a big fly is smaller than a small mouse). We cannot just say a glass is fragile. We have to say *how* fragile it is if we want to discuss whether it will break when we drop it. This gets back to *typicality*. When we call an object fragile we mean that it is liable to break when handled in ways that we would normally handle objects of that type; so 'it is fragile' can be taken to mean something like 'handle with extra care.'

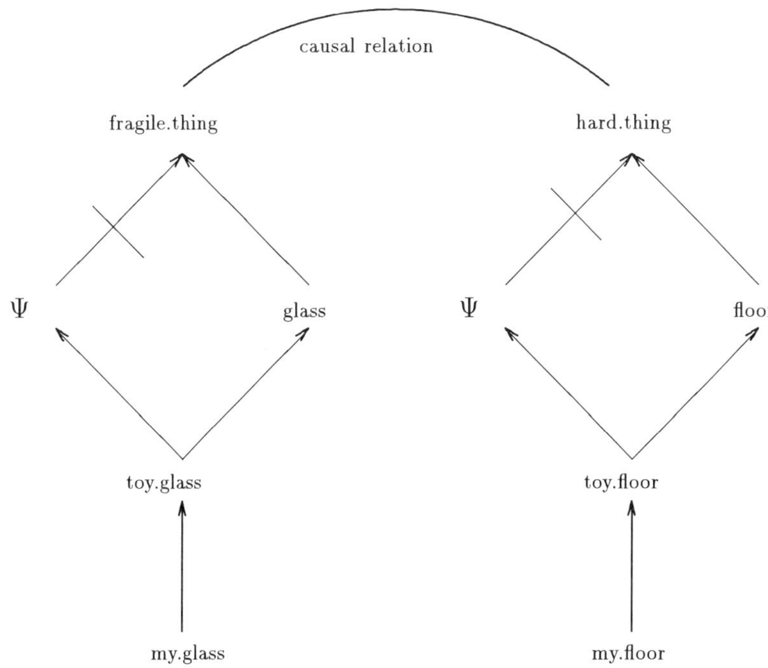

Figure 7.6: Exceptional state due to the exceptional inheritance classes

To reason by causation, causal relation checks the inheritance structure classes and the subclasses behind it. If this class has an explicit property then this property will override any implicit (default) property, otherwise exceptional properties will override the default properties. In other words, causal reasoning depends on the exceptional inheritance structures and follows its is-a and not-isa relations.

We can summarize as follows:

1.With causal relation, the notion of properties (states), like 'hard' and 'fragile', is relative.

2.The unexpected states (effects) of an action are due to the exceptional inheritance classes (Ψ) that are located in the base of a causal relation.

7.4.2 A Semi-formal Introduction to Inheritance-Causal Relation

Accordingly, we can formalize a causal rule using the notions and the mechanisms that have been used with EIR in Chapter 5 as follows:

$\mathcal{C}(x,y)\bullet[\mathcal{P}(x) \wedge \mathcal{Q}(y)] \odot [[\neg\zeta(x) \wedge \neg\beta(x)] \wedge [\neg\zeta(y) \wedge \neg\beta(y)] \wedge \chi] \rightarrow \mathcal{E}(x,y)$

where

- $\mathcal{C}$ = Action or Cause between x and y.
- $\bullet$ = Applied to.
- $\mathcal{P}$ = Property of the individual x that is inherited via its related typical class within the inheritance network.
- $\mathcal{Q}$ = Property of the individual y that is inherited via its related typical class within the inheritance network.
- $\neg\zeta(x)$ = x is not an inheritable class, its properties related to the other exceptional class Ψ.
- $\neg\beta(x)$ = x is an acquired class, its properties are the explicit ones.
- χ = background conditions (or other circumstances).
- $\mathcal{E}$ = the state (effect) of x or y or both after an action is applied.

For example, 'throwing my.glass on my.floor cause the breaking of my.glass' could be represented as follows:

Let

- C = the input action (throwing).
- E = the output effect (breaking).
- my.glass is an instance related to the typical.class 'glass' and the default property 'fragile.thing'.
- my.floor is an instance related the typical.class 'floor' with the default property 'hard.thing'.

- $\neg\zeta$ my.glass is not (for instance) a plastic.glass. Same for 'my.floor'.
- $\neg\beta$ my.glass is not (for instance) a 'fly.glass'. Same for 'my.floor'.
- χ is the general background conditions that is not related in specific to my.glass or my.floor (such as the distance that we threw the glass from or the force with which the glass is thrown, etc).

Therefore applying inheritance-causal relation will produce:
throwing(my.glass, my.floor) $\bullet$
[fragile.thing(my.glass) $\wedge$ hard.thing(my.floor)] $\odot$
$[[\neg\zeta(my.glass) \wedge \neg\beta(my.glass)] \wedge [\neg\zeta(my.floor) \wedge \neg\beta(my.floor)] \wedge \chi]$
$\rightarrow$ breaking(my.floor, my.glass).

7.5 Exceptional Inheritance-Causal Algorithm

In order to specify an EIR-causal algorithm, we refer to the notation described in Chapter 5 and Figure 7.7. We start by an input causal statement. Analyze such statement for its action (implicit or explicit), instances, related superclasses, and states (properties). Relate the top-level properties by the defined action. Apply the EIR algorithm on the two inheritance structures to produce the expected effect.

From this algorithm we can specify the generality for any causal statement after defining the inheritance structures behind it.

7.6 Script-Story Understanding

Moreover, the EIR approach also can be generalized to formalize a causal chain (Gan, 1993), for example, used in story comprehension as a script filling the gaps between isolated events in the story (Schank, 1986). Consider the story F as a sequence of isolated events $e_1, e_2, \ldots, e_n$. The comprehension system must try to make these isolated events causally well connected. Causal relations between subsequent events play a very important role in the understanding process (Schank and Abelson, 1977). We distinguish between two types of causation, namely implicit causation and explicit causation. To put it simpler, we take an example to

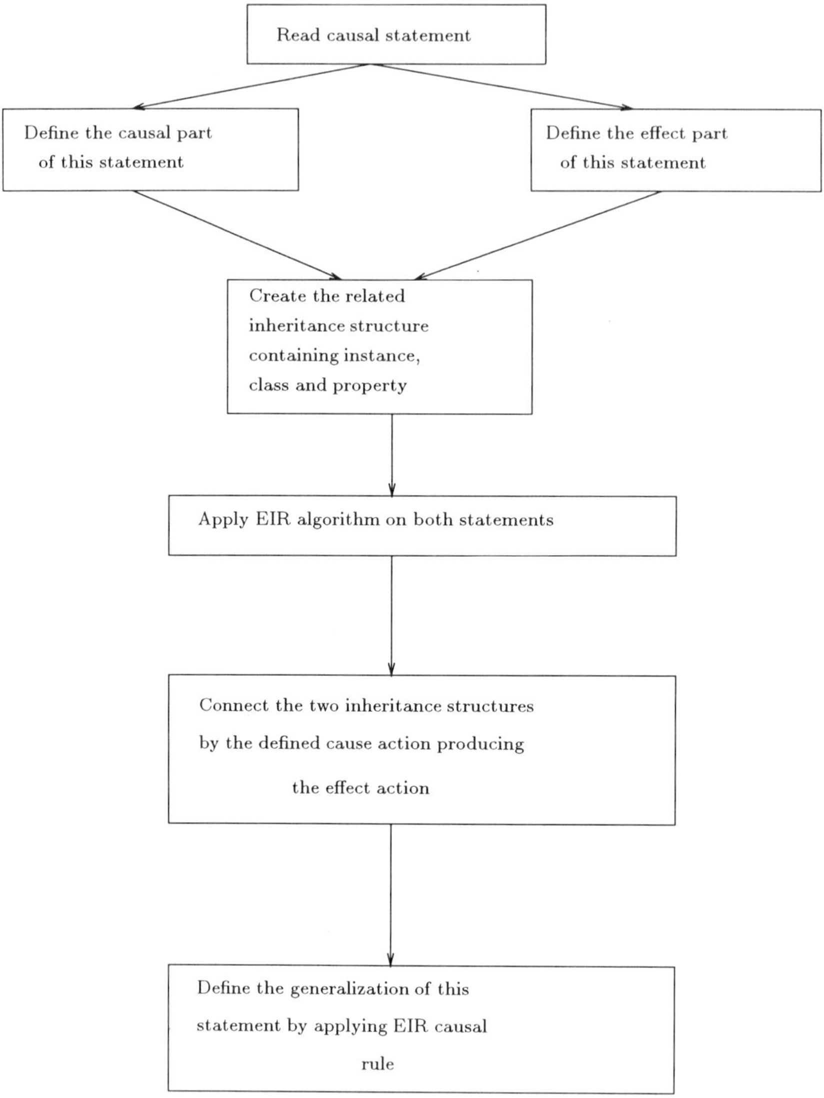

Figure 7.7: EIR-causal algorithm.

specify how the EIR works in script-based story understanding. The typical story is:

Story 1 *John went to the restaurant (e_1). He ordered a dish of lobster (e_2). He ate it (e_3) and paid the check (e_4). He left (e_5).*

Here, the story has five isolated events. When people hear this story, they remember their past experiences about what normally happened in the restaurant: You entered the restaurant, found a seat to sit down, looked at the menu to find tasteful dishes, ordered these dishes, paid the check, and finally left. All these actions have repeatedly taken place in such situations so that people are likely to follow such a script to fill in gaps between isolated events. In practice, the story arrives step by step. Therefore when people heard the first event (*John went to the restaurant*) they may predict that the next actions normally would follow the script. Behind the script, which described a sequence of actions, there is *implicit causality* between adjoint events. For instance, why did John go to the restaurant? The reason may be that he was hungry and he had enough money among other reasons (χ). After he entered the restaurant, causally he had to find a seat to sit down, and then he had to order, eat, pay, and leave. The *explicit causality* here is that there are seats to sit down in, there is a waiter to serve and there is money to pay – causality between the actions.

Now we extract the actions to represent the events, so that we can easily create a corresponding inheritance network, shown in Figure 7.8. Let other implicit causalities correspond to inheritable classes (they refer to an implicit deviation from the normal script) and explicit causalities to acquired classes (they refer to an explicit deviation from the normal script). According to EIR2, we have the following inference rules:

$$Entered(John) \odot [\{\neg\zeta_0(John) \wedge \neg\beta_0(John)\} \wedge \chi_0] \rightarrow Sat.down(John),$$

$$Sat.down(John) \odot [\{\neg\zeta_1(John) \wedge \neg\beta_1(John)\} \wedge \chi_1] \rightarrow Looked.up.menu(John)$$

$$Looked.up.menu(John) \odot [\{\neg\zeta_2(John) \wedge \neg\beta_2(John)\} \wedge \chi_2] \rightarrow Ordered(John)$$

$$Ordered(John) \odot [\{\neg\zeta_3(John) \wedge \neg\beta_3(John)\} \wedge \chi_3] \rightarrow Ate(John)$$

$$Ate(John) \odot [\{\neg\zeta_4(John) \wedge \neg\beta_4(John)\} \wedge \chi_4] \rightarrow Paid(John)$$

$$Paid(John) \odot [\{\neg\zeta_5(John) \wedge \neg\beta_5(John)\} \wedge \chi_5] \rightarrow Left(John)$$

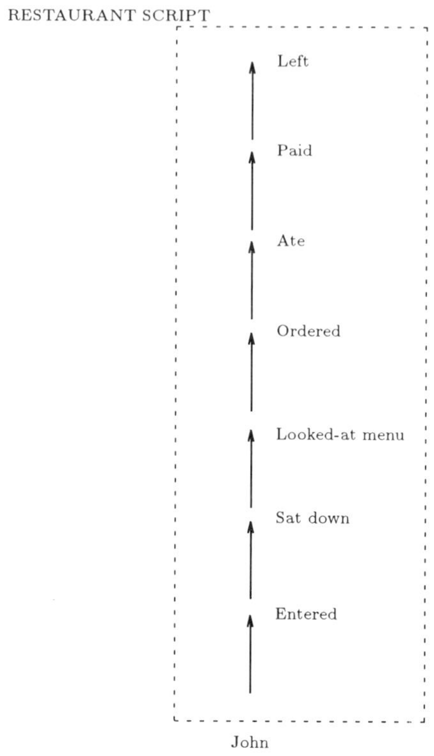

Figure 7.8: The translated inheritance network of Story 1.

where ζ_i is an implicit exception, β_i is an explicit causality, and χ_i is a background condition. Fortunately, in this story there is indeed neither explicit causality to deviate from the normal way nor other implicit causality between the actions except the original causal explanation under which the sequence of actions happened. Namely the conclusions can be explained as follows: John went to the restaurant, provided that there is no exception to that, John found a seat to sit down, and there is no other implicit causality different from his attempt to sit down (there is an implicit causality that when one entered a restaurant, he had to sit down to eat something), then John found a seat to sit down: John sat down. Provided that there is no exception to that, he looked at the menu (and there is no other implicit causality different from that, he normally had to look at the menu after he sat down), then he looked at the menu, and so on.

Story 2 *John went to the restaurant* (e_1)*. He spoke to the manager* (e_2)*. He did not pay* (e_3)*. He left* (e_4)*.*

This story is quite different from the previous one. Here, although John went to the restaurant, his aim was not to eat something, but to talk to the manager. The translated inheritance network is shown in Figure 7.9. Obviously, the script is not valid when one tries to understand this story because the actions are no longer happening as described in the script.

According to EIR2, we have

$$Entered(John) \odot [\{\neg\zeta_0(John) \wedge \neg\beta_0(John)\} \wedge \chi_0] \rightarrow Sat.down(John),$$

where the causal exception ζ_0 is that John spoke to the manager, and there is no other implicit causality different from the normal explanation, which said that after John entered the restaurant he had to find a seat to sit down, eat and pay. Furthermore, following the normal script we have:

$$Entered(John) \odot \neg Spoke(John) \rightarrow Paid(John)$$

But the story told us something different

$$\odot Spoke(John) \rightarrow \neg Paid(John)$$

Now it is easy to identify the contradiction:

$$Entered(John) \odot Spoke(John) \rightarrow \otimes Paid(John)$$

According to the EIR, there exists a special class Ψ, such that

$$Spoke \rightarrow \Psi, \Psi \not\rightarrow Paid,$$

namely, there is something behind that John spoke to the manager about, which made John not pay. By EIR6, we have

$$\Psi \bowtie Ate$$

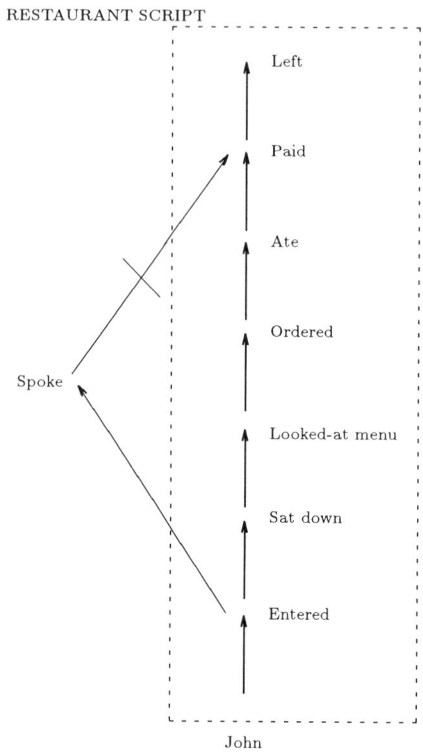

Figure 7.9: The translated inheritance network of Story 2.

Now we obtain the conclusion that John did not sit down, hence did not look at the menu, order dishes or eat something from the following:

$$Entered(John) \odot \{Spoke(John) \wedge \neg\beta(John)\} \rightarrow \neg Paid(John)$$

Story 3 *John went to the restaurant (e_1). He found there is no lobster left (e_2). He did not pay (e_3). He left (e_4).*

This story does not follow the script, either. The reason is different from Story 2 – John's aim is to eat something in the restaurant but

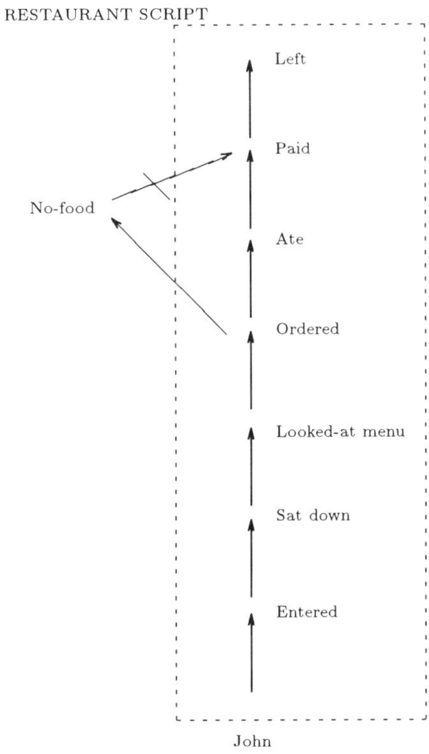

Figure 7.10: The translated inheritance network of Story 3.

unfortunately there is no preferred food left for him. The translated inheritance network is shown in Figure 7.10.

In such a case, we treat this reason as an explicit causality obviously different from the normal one, which said after John ordered a dish he would eat it. So we have $\beta = Nofood(John)$ such that

$$Ordered(John) \odot [\{\neg\zeta_3(John) \wedge \neg Nofood(John)\} \wedge \chi_3] \rightarrow Paid(John)$$
$$\odot Nofood(John) \rightarrow \neg Paid(John).$$

Similar to the previous analysis, we obtain that John did not pay anything and left.

We analyze the restaurant examples to demonstrate that EIR theory can be used for formalizing and handling more complicated situations. In script-based story understanding, we catch the idea that there is explicit causality and implicit causality behind normal action sequences.

Chapter 8

Application: Analogical Reasoning and EIR

8.1 Introduction

Analogical reasoning is one of the central inference mechanisms in human cognition and a powerful computational mechanism. The bulk of human problem solving by analogy takes place in problem spaces that are either well known or vary only slightly from familiar situations.

The need for analogy in AI arises when we acquire knowledge to create a new rule or a new fact. It assists the transformation of knowledge from a past situation that bears a strong similarity to the problem under investigation to a new one. The rules or facts thus created will be transformed to the new situation and stored for future unrecognized problems. *Reasoning* takes place when analogy tries to solve a problem about one domain or situation given a similar stored situation.

The aim of this chapter is to develop a formal framework for analogical reasoning by using inheritance reasoning and casual relations (explained in Chapter 7). This formula will help a user to define new relations for a target, problem, or space easily. This is explained in detail in Section 8.5.

There is argument as to the role of analogy in cognition. A common view is that analogical reasoning is powerful because it allows us to bring knowledge, frequently from a well-understood domain (*the base*), to explain a less understood domain (*the target*), with a dramatically reduced reasoning effort.

The importance of *analogical reasoning* has been acknowledged by many researchers. For example, the use of analogy in problem solving and learning was proposed by, among others, Carbonell (Carbonell and Michalski, 1983a; Carbonell and Michalski, 1983b; Carbonell, 1986), Winston (Winston, 1980), Falkenhainer *et al.* (Falkenhainer and Forbus, 1989), Kedar-Cabelli (Kedar-Cabelli, 1988), Burstein (Burstein, 1986) and Kodratoff (Kodratoff, 1988)[1]. Winston (Winston, 1984) demonstrated the importance of past experience as being one element in a hypothetical intelligent machine. Winston stressed the importance of past experience in learning, where people use their experiences to understand new subjects, such as medicine, law, and physics, for example. Carbonell (Carbonell and Michalski, 1983b) introduced transformational analogy between a base domain and a target domain. In other words, a target problem is solved by searching in a space of possible solutions in a base domain and then transferring the solution (by an object-to-object mapping) to a target domain. Psychological theories play an important part in analogical reasoning studies, too. Rumelhart and Norman (Rumelhart, 1981), and Gentner (Gentner, 1983) provided examples of such theories. Gentner (Gentner, 1982) used well-defined relations in a rich domain to understand and explain relations that exist in a poor domain. Most of these analogical reasoning models have been developed to form a kind of partial pattern matching. The mechanisms of these models are based on the assumption that the best partial match can be found by accumulating facts and rules for possible object-to-object mappings between two domains.

8.2 The Role of Analogical Reasoning in AI

Analogical reasoning is the cornerstone of artificial intelligence research. In AI, researchers try to capture this role in an 'intelligent machine' by formalizing people's ability to understand a new situation and in

[1]A detailed survey of the computational approaches to analogical reasoning is found in, among others, (Ellman, 1989), (Hall, 1989), (Kodratoff, 1988), and (Kedar-Cabelli, 1988).

solving the new problem. In other words, formalizing people's abilities in understanding new situations by comparing their relationships to old ones and solving new problems by using previously solved analogous problems.

Reasoning, in general, starts from the point, when we acquire knowledge, and the real world around us supplies us with an incomplete, inconsistent picture in addition to the exceptional information. According to this and to reach a reasonable decision when trying to solve a new problem, two approaches are adopted. The first is to stay within the problem's domain without using any knowledge from outside it. This means that in order to reason within the same domain (to expand or explain the domain's knowledge), a specific kind of reasoning, such as deductive, default, or any other kind of nonmonotonic reasoning is used. Inheritance reasoning, in some sense, may be seen as an application of such an approach. The other approach is to use knowledge from other domains. This is done by searching for a similar, but richer domain, which will be used as a knowledge supplier. In this case, analogical reasoning could be seen as one instance of such an approach. Either method benefits from well-defined inheritance structures and the use of EIR. AI's concern with creating practical reasoning machines has led to the examination of different aspects of commonsense reasoning, such as deductive, inductive, and nonmonotonic reasoning. To make analogy applicable, a combined version of nonmonotonic reasoning and analogical reasoning is needed. This is because nonmonotonic reasoning is a kind of commonsense reasoning, in the sense that it may reach a conclusion in the absence of complete knowledge, whereas analogical reasoning is an inductive reasoning process – it reaches conclusions using other, rich domain knowledge. Thus, analogical reasoning may be seen as a special kind of nonmonotonic reasoning.

8.3 Reviewed Literature

ANALOGY (Evans, 1968) was one of the earliest systems created to solve traditional geometric problems, in this case using analogies between different combined geometric shapes.

Figure 8.1 shows a simple analogical reasoning problem: Given that an A is related to B as C is related to X, find X. ANALOGY worked

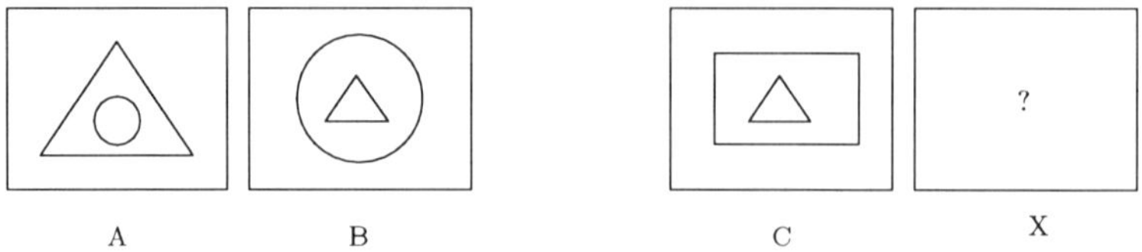

Figure 8.1: Reasoning by analogy.

on the assumption that the best matching between two situations will be done by accumulating evidence from the changes between relations in the base domain. ANALOGY's inputs are a symbolic description of the geometric shapes and the relationships between them. An *elaboration* to the base domain, by inferring all the possible relationships from the input shapes, was then done. Rules were governed by the possible transformations from one shape to another in the base. A generalization (extension) rule mapped the set of transformation rules in the base to the target. Finally, and in order to choose the best rules, a *justification* step was done. *Best rule* in Evans' system tends to mean the longest (e.g., in specification) rule.

Kling (Kling, 1971) created a program for constraining axioms. The axioms assist in proving a new theorem by analogy with a similar proof in an analogous theorem. The input consists of the target theorem to be proved, an analogous base theorem and its proof in this case.

Winston (Winston, 1975; Winston, 1980; Winston, 1984) used similarities between situations such as stories and physical experiments, in learning new concepts by analogy. Winston argued that much of human thinking is done by analogy. When a person faces a new situation, he/she tries to recall another situation that was similar in some aspects. Then he or she matches the two situations with each other. This matching process allows learning of new facts (or rules) about the new situation. Winston stressed that important relations are learned either from a teacher or by past experience. A situation can be described using properties, actions, and other relations. One situation is similar to another if the important relations can be replaced in correspondence. In addition, he proposed transferring causal networks between domains: A causal structure in a well understood situation is likely to provide information to a poorly understood situation. Finally, Winston asked 'How is it possible to know if some relation holds in one situation, given that the situation is analogous to another, well understood situation?' He answered that the constraint relations in the well understood situation suggest the right relation to check and the right question to ask.

Like Winston, many other people believe that causality has an important role to play in learning and understanding. Schank (Schank, 1975) claimed that people can cause and prevent actions. People can also persuade and dissuade, order, forbid, and force. In that sense, causality can be used in defining similarity relations in understanding two different domains.

In a different approach, Structure-Mapping Theory (SMT); (Gentner, 1983), the comparisons between two domain structures form the basis of the rules that are used to import a descriptive structure (relations, attributes, and causal chains) from a base domain to a target domain. SMT's central idea is that predicate structure can be transferred from one domain to another with different objects and different surface appearances (Gentner, 1983; Falkenhainer and Forbus, 1989; Forbus, 1986). One example used by Gentner to illustrate her theory is the well-known analogy of describing atoms as being like a miniature solar system. The aim is to augment the representation of the base by mapping the set of predicates to the target domain. Predicates used with SMT have a kind of order: A causal relation is an example of predicate of higher-order relations that have a priority over low-order relations when we match two structures[2]. SMT considers the *systematicity principle* as the central principle to analogy (Falkenhainer and Forbus, 1989). The systematicity principle is the number of relations that can be mapped between two domains. This quantity determines the best possible match. For example, SMT uses the causal relation (higher order relation) behind the statement 'heat flows through a bar' to explain its analogy 'water flows through a pipe' [3]. SMT ignores the appearance of some attributes,

---

[2]Gentner defined proposition's orders as follows: Constants have order zero. A proposition's order equals the maximum of the orders of its arguments plus one. Causal relations are of order two (higher-order) because they contain at least two propositions of different arguments (Forbus, 1986).

[3]See Figure 8.3 in Section 8.4.

such as 'the water is wet' or 'the bar is a special material.'

Burstein (Burstein, 1986) created CARL, a system that learns concepts by forming analogies. Burstein illustrated his method by saying, about a computer program written in the language BASIC':

> To illustrate the concepts of variable, imagine that there are 26 little boxes inside the computer. Each box can contain one number at any one time. The analogy shown in [Figure 8.2] can be paraphrased as follows: 'A *variable* is like a *box* in that numbers can be *inside* variables similar to the way objects can be inside boxes.' (p. 353).

A 3	B	C	D	E 12	F	G	H	I
J	K	L 7	M	N	O	P	Q	R
S	T	U	V	W	X 5	Y	Z	

Figure 8.2: Numbers inside variables.

Through the analogies between boxes and variables, CARL learns about the semantics of assignment statements for the BASIC language. Like a real student CARL uses trial and error as part of the learning process. In Burstein's own words,

> It was often the errors made by these students that provided the most useful insights into analogical reasoning process.

Both the base and the target concepts are assumed to be available to CARL as inputs. Naturally, CARL, as other systems, had some limitations to the kinds of relations that could be used in the mapping process, leaving other relations to proceed, such as causal relations.

Carbonell's proposal is a model of transformational analogy using *Means-Ends Analysis* (MEA) (Carbonell and Michalski, 1983a). MEA refers to the problem-solving method which consists of selecting actions that reduce known differences between the current situation and the desired state (Carbonell and Michalski, 1983a). In other words, a new problem could be solved by using a solution from a space of possible solutions; that is, a space in which states contain a sequence of operator invocations in the original space. Carbonell called this space a transformational space, or *T-space*. 'When facing a new problem situation, a person is reminded of past situation(s) that bear strong similarity to the present problem at different level of abstraction' (Carbonell and Michalski, 1983a). The following states with different kinds of information will be recognized twice, first in the base domain and second in the analogical target domain:

- A given situation (information) or an *initial state.*
- A goal situation or a *final state.*
- The transformation operators or *T-operators* that are responsible for transformation from an initial state to a final state.

MEA is applied to the base situation in order to adjust (insert, modify, delete, and reorder) problem-solving operators. For example (given in (Carbonell and Michalski, 1983a), John is located in Pittsburgh and must travel to New York City. After a phone call, John discovered that all airline flights were booked. John never took the intercity train before. John knows that it is possible to do so by using his transformational analogy as follows:

- Initial state: Being reminded of taking an airplane (which includes enough money, advance reservation, ticket to be issued, etc).
- Final state: John boards the train.
- T-operators: Substituting train travel for air travel which replaces 'airport' by 'train station', 'airline ticket' by 'train ticket', and so on.

Carbonell did not formulate a structure in the base that could be used in order to apply the transformational analogy to the target.

To address the shortcomings of the transformational analogy approach Carbonell (Carbonell, 1986) presented the *derivational analogy*. The problem with transformational analogy is that different situations may share the same initial state but not the same solution (Kedar-Cabelli, 1988). Derivational analogy uses analogical mapping between the base and the target by addressing the *plan* or the *decision steps* that humans use in solving a problem. For example, it should be easy to program a quicksort in LISP for somebody who has an implementation to quicksort in Pascal, but a Pascal implementation of such a program looks different from a LISP implementation. Transferring corresponding steps from the Pascal program into LISP is not the right method to produce a LISP program (Carbonell, 1986). What is needed here is the algorithm that has been used in implementing quicksort in Pascal. The derivational analogy follows the reasoning process in constructing past solutions and considers whether they are still appropriate in the new situation or need to be modified in light of the differences between them (Carbonell, 1986).

Kedar-Cabelli (Kedar-Cabelli, 1985; Kedar-Cabelli, 1990) presented Purpose-Directed Analogy, an approach that uses explicit knowledge of purpose as input to create a precise, relevant causal network. His approach criticizes the Gentner (Gentner, 1983), Burstein (Burstein, 1986), and Winston (Winston, 1980) approaches, in that they use and map all relevant and irrelevant causal relations between two domains without any consideration of the purpose behind the mapping purpose. In other words, Kedar-Cabelli uses 'purpose' as an additional constraint in any possible mapping between two domains (Kedar-Cabelli, 1988).

8.4 Analogical Reasoning Revisited

The aim of this section is to examine one analogical reasoning model , namely SMT (Gentner, 1982). SMT describes the set of implicit constraints by which people interpret analogy and similarity. SMT's central idea is that an analogy is a mapping of knowledge from one domain (the base) into another (the target), which conveys the information that a system of relations known to hold in the base also holds in the target

(Falkenhainer and Forbus, 1989; Forbus, 1986).

Gentner's goal is to augment the representation of target knowledge by mapping over relevant aspects of the base facts. Gentner's learning concept can be explained as follows. After introducing base and target domains, a *retrieval* process that places relations and entities of base and target in correspondence will take place as a first step. The second component of SMT is *elaboration*, which refers to the additional attributes, relations, and causal chains that could be derived from the base domain to help in defining similar relations in the target domain. The third step is *mapping*, which prefers high-order relations (such as causal relations in the base domain) over independent relations to be mapped to the target. The mapping process ignores the irrelevant relations that are outside the meaning of the analogy. The fourth and final stage of Gentner's SMT is *justification*, or modification the attributes and the relations that are mapped over relevant aspects of the base in the target domain.

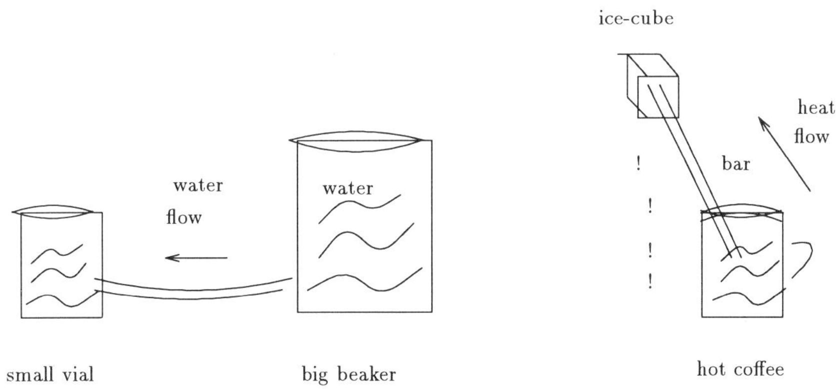

Figure 8.3: Water flow is similar to heat flow.

Figure 8.3 (Falkenhainer and Forbus, 1989; Forbus, 1986) demonstrates such an analogy. Consider the water flow and the heat flow situations. These situations are thought to be, under Gentner's SMT, analogous because they share *the complex relationship* known as *flow*. Falkenhainer et al. (1989) wrote:

> 'We prefer to ignore the appearance and even specific defining properties of the object, such as the fact that water and coffee are both liquids' (p. 3).

Two situations are thought to be analogous by highlighting the appearance of the structure's properties. What is not being questioned in this chapter is the claim that analogy is about relations rather than simple features. What is being questioned here is this: What is the structure type that embodies these relations? What is the semantics behind such structures? What is the reasoning procedure that we need to follow in the mapping process? These questions are addressed in the remainder of this chapter.

8.5 Analogical Inheritance Reasoning

The aim of this section is to study the relation between analogical reasoning and inheritance structures. This relation has three aspects:

1.Assume we have the following analogy:

Water flow through a pipe is similar to heat flow through a bar.

We may analyze this analogy between the two structures as shown in Figure 8.4. These two structures can be considered as inheritance structures containing the individuals 'water and heat', the classes 'liquid and energy' and the property 'flow', respectively. The classes 'liquid and energy' share the same top level property 'flow'. In other words, we are using inheritance structures as a base to explain the analogy between two domains.

Hence, starting from the two situations, we can define their inheritance structures. Using these inheritance structures allows us to use the top-level properties. This provides an analogical relation <u>between</u> the two inheritance structures.

2.After considering the previous analogy and because they share the same top-level property as one structure, we may combine the two inheritance structures as one. For instance (see Figure 8.5) we apply 'high pressure' on 'water' in a similar way to applying 'high temperature' on 'drink.coffee'. On the one hand, and because 'drink.coffee' and 'water' are two instances related to the class 'liquid', they inherit its property *flow* by default. On the other hand, whereas 'high pressure' is an in-

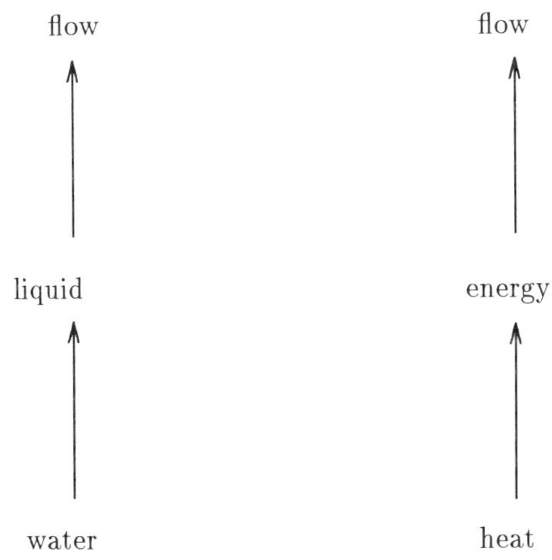

Figure 8.4: Water flow is similar to heat flow structures

stance of the class 'pressure', 'high temperature' is an instance of the class 'temperature'. Both of them are related to the superclass 'energy' which inherits the property 'flow' by default. Because these classes share the same top-level (default) property, this gives us the chance to replace an object with another within the same inheritance structure. In other words, this may be considered as a heuristic method in defining new objects with an inheritance structure. That is, we use the top-level inheritance properties to identify the alternative objects or classes in order to use them to reason by analogy. We use the complex relationship 'flow' (Gentner, 1982) as a top-level default property in the two inheritance structures.

Thus, if we define a set of classes that gather together a number of similar subclasses and individuals, then we will be able to replace any one of these subclasses by another to produce the same property.

We call the approach that uses top-level properties to define analogical classes and instances the top-down approach. That is because we use the shared properties in defining the different classes and individuals

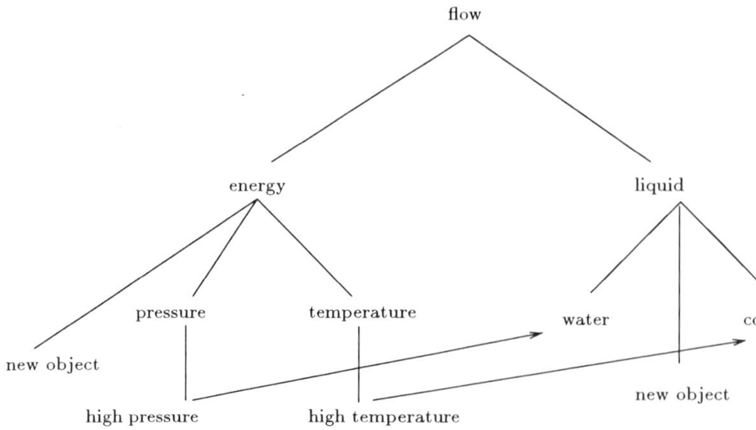

Figure 8.5: Combine two inheritance structures sharing the same top-level property.

within these inheritance structures.

3.As defined in Chapter 7, we may define a causal relation as a relation between the top-level properties of two inheritance structures. From an analogical reasoning point of view, we may use this relation in defining other alternative classes in two analogical domains.

Figure 8.6 demonstrates the role of causal relations and inheritance structures in defining the analogical parts between two domains.

Consider Figure 8.7. We can adopt the proposed analogical reasoning mechanism as follows: The following are given:

Object-1 ... Object-n share the same top level inheritance Property ZZ,

Object-11 ... Object-nn share the same top level inheritance Property KK,

Object-1 has an effect on Object-11 via the defined causal relation between ZZ and KK.

We say that Object-1 in the first domain has a chance to play the same effect (by analogy) on Object-jj in the second domain. At the same time there is a chance for any other class, such as object-i, in the first

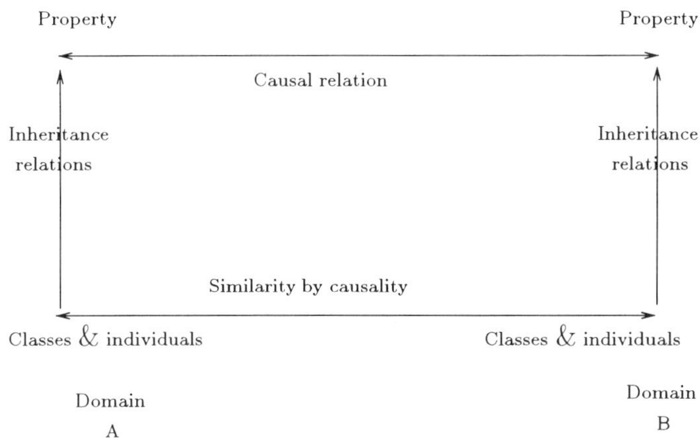

Figure 8.6: Analogical reasoning using inheritance structures and causation.

domain to play the same role (by analogy) on any other class in the second domain, such as object-jj. This is the analogical reasoning that relates any two inheritance structures by causal relation.

In summary, inheritance structures can play a significant role in defining or explaining the analogical parts using top-level properties between two analogical domains. Causal relations between properties of inheritance structures have a role as well.

8.6 EIR and Analogical Reasoning

In Chapter 5 we defined EIR, an exceptional inheritance reasoner with a clear semantics. In this section, EIR is used to provide a base structure to formalize analogical reasoning. To do so, let us demonstrate the relationship between causal reasoning and analogical reasoning using exceptional inheritance reasoning as follows. Consider Figure 8.8 and the following causal relation:

Throwing 'my.glass' on 'my.floor' causes the breaking of 'my.glass'.

We may formalize such a statement (as demonstrated in Chapter 7)

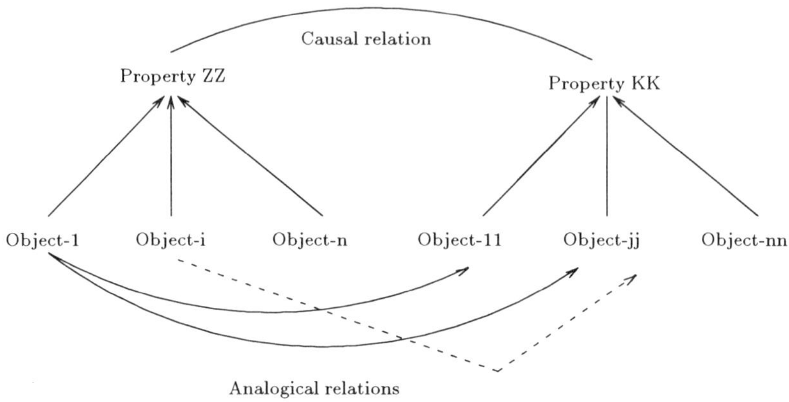

Figure 8.7: Analogical inheritance reasoning.

as follows:

throwing(my.glass, my.floor) $\bullet$
[fragile.thing(my.glass) $\wedge$ hard.thing(my.floor)] $\odot$
$[[\neg\zeta(my.glass) \wedge \neg\beta(my.glass)] \wedge [\neg\zeta(my.floor) \wedge \neg\beta(my.floor)] \wedge \chi]$
$\rightarrow$ breaking(my.floor, my.glass).

This causal relation leads to the following general rule: Applying an action on an object will cause an effect provided that the following hold:

- 'Object-1' is an instance of 'Class-1' with the default property Property-1'.

- 'Object-2' is an instance of 'Class-2' with the default property 'Property-2'. And

provided that

- 'Object-1' and 'Object-2' are not classes of exceptional properties.

- 'Object-1' and 'Object-2' are not classes of acquired properties.

- All other conditions are known (χ).

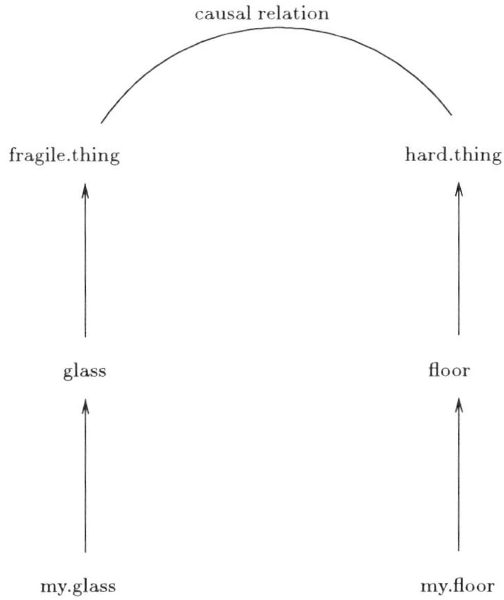

Figure 8.8: The inheritance-based analogical structure.

So for the given example: Object-1 = my.glass, Class-1 = glass, Property-1 = fragile.thing, Object-2 = my.floor, Class-2 = floor and Property-2 = hard.thing, the action is 'throwing' and the effect would be 'breaking'.

Now let us introduce the exceptional case, in which 'my.glass' is not a 'fragile.thing' (it happens to be an exceptional class related to the superclass 'glass'). In order to formalize such a case, and because the individual 'my.glass' is an unfragile.thing, applying EIR will create a different extension.

Consider the following primitive paths (see Figure 8.9).

$$my.glass \rightarrow elastic.glass, \qquad elastic.glass \rightarrow glass,$$

$$elastic.glass \not\rightarrow fragile.thing.$$

and

$$my.floor \rightarrow soft.floor, \qquad soft.floor \rightarrow floor,$$

$$soft.floor \not\rightarrow hard.thing.$$

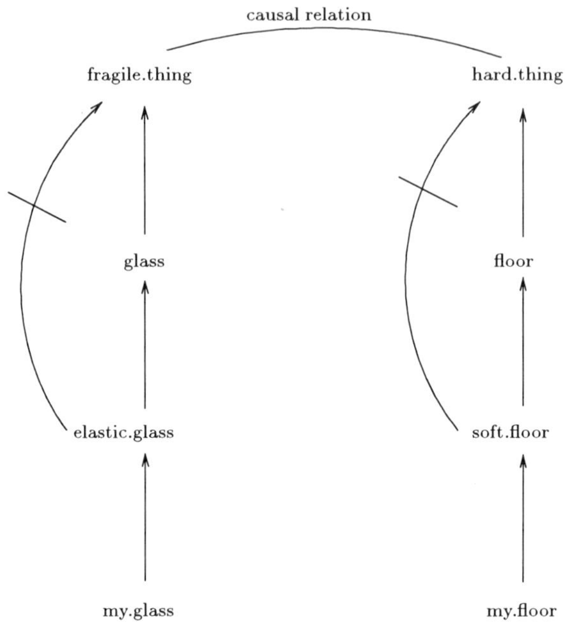

Figure 8.9: EIR-based analogical structure.

From these paths, and in order to reach the effect of the action 'throwing', we need to check whether there is any contradiction between the assertions above by using PR2 and PR3 of Chapter 5 as follows:

$$Glass(my.glass) \odot \neg elastic.glass(my.glass) \rightarrow fragile.thing(my.glass)$$

$$\odot elastic.glass(my.glass) \rightarrow \neg fragile.thing(my.glass)$$

and the same for the other domain,

$$floor(my.floor) \odot \neg soft.floor(my.floor) \rightarrow hard.thing(my.floor)$$

$$\odot soft.floor(my.floor) \rightarrow \neg hard.thing(my.floor)$$

These primitive relations, identify the contradiction

$$glass(my.glass) \odot elastic.glass(my.glass) \rightarrow \otimes fragile.thing(my.glass)$$

and

$$floor(my.floor) \odot soft.floor(my.floor) \rightarrow \otimes hard.thing(my.floor)$$

The exceptions ($\odot$ *elastic.glass*$(my.glass)$ and $\odot$ $soft.floor(my.floor)$) and the default conclusions

($\otimes$ $fragile.thing(my.glass)$ and $\otimes$ $hard.thing(my.floor)$) will be used to locate the exceptional class Ψ, which will assist in explaining the exceptional properties of the individuals my.glass and my.floor (see Figure 8.10).

The structure in Figure 8.10 demonstrates the following: Because the EIR and causal relations between those structures are defined, *similar classes* that share the same top-level properties, using analogical reasoning, may also be defined. For instance, we may replace the typical class 'floor' by the typical class 'Iron' because both of these classes, as far as inheritance is concerned, share the same top-level property 'hard.thing'. The same procedure may be applied on the second inheritance structure with the property 'fragile.thing'.

In summary, as far as inheritance is concerned, EIR can assist in defining the inheritance structures behind an analogical statement. This is done by defining the shared top-level property of these structures. On the other hand, using a causal relation as a relation between the top-level properties of two inheritance structures, we may define the alternative classes within each of the inheritance structures. Such an approach will save a reasonable amount of reasoning time, which is one of the aims of AI research.

What must be stressed here is that this chapter has been concerned only with demonstrating the potential use of EIR to analogical reasoning. A full evaluation of the potential will depend on implementing and developing reasoning systems. Although implementation is a necessary step in the evaluation of EIR's potential in this problem domain, the aim of this chapter is different. Our task has been to strengthen the claim that EIR represents a novel approach to commonsense and practical reasoning, and this has been achieved by demonstrating how EIR's

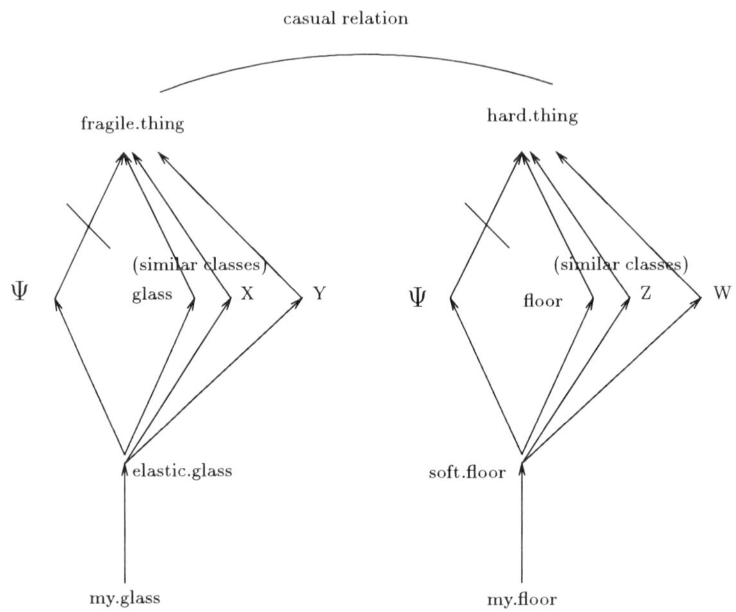

Figure 8.10: EIR application of analogical reasoning

structures can be applied to analogical reasoning, as a paper exercise. The main concern of the book as a whole is with EIR as an AI reasoning tool. Together with the previous chapter (as well as Section 7.6 of Chapter 7 as another application), these two chapters should be interpreted as *further* support for EIR as introduced and defined in Chapters 1, 4, 5 and 6 and not as the *only* support. These chapters also indicate where future work on EIR may be concentrated.

Chapter 9

Conclusion

9.1 What has been Achieved

In this book, I have presented the general case for the potential benefits of intelligent computer systems. Appropriate representations of commonsense knowledge are, I have claimed, crucial to the long-term goal of devising such systems. And a major part of this prerequisite knowledge representation problem is a proper treatment of typicality, exceptions, incomplete knowledge, and knowledge revision within an overall hierarchical organizational structure.

As a basis for our treatment of these problems, I have introduced a new theory of inheritance in which I tackled a clash of intuitions concerning inheritance structures. Central to the book has been the identification of a general formalization principle that aids the resolving of redundancy as well as ambiguity. I have treated redundancy as a problem of revising new and true information in the light of originally true yet different information. Within an inheritance structure, the original and the new information (represented as paths in a semantic network) emanate from the same start point to converge at the same end point.

In Chapter 5, we provided the basic foundations of *Exception-based Inheritance Reasoning*, or EIR. EIR involves a novel interpretation of *not-isa* links: It runs as follows:

A typical class is one that passes information to its subclasses or instances by default, as is the case with standard class specialization. If the inheritance structure consists solely of typical classes linked by *is-a* links, there is no problem in identifying the typical properties to be inherited by instances. But a class may also be typical in virtue of providing exceptions to more general typical classes. The approach allows exceptional information that is nevertheless inheritable to be encapsulated in a class, called an *exceptional class.*

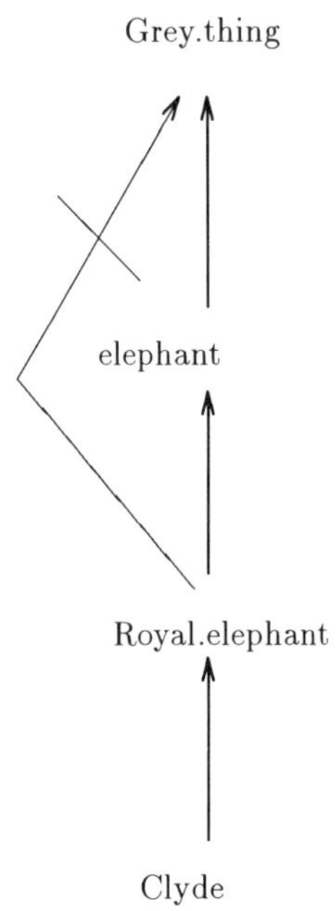

Figure 9.1: Exceptional Inheritance Reasoning

Consider the 'Clyde' structure (Figure 9.1). Clyde, although an elephant (through two *is-a* links), is a Royal.elephant that typically is not a grey.thing. *Elephant* in this example is a typical class in that elephants typically are grey things. *Royal.elephant*, though, contains an exception to the typical class to which it belongs, (*elephant*).

EIR's interpretation of this is that *Royal.elephant* must have inherited the property of typical nongreyness from an exceptional class for which nongreyness is typical. Information that is exceptional to the typical class and that belongs to the exceptional class must be inherited as typical by subclasses and instances of the exceptional class, unless overridden.

The reasoning behind this interpretation is as follows. In the current literature the *is-a* links are used to specify superclass relationships as well as to infer inheritance relationships in the typical class case. In exceptional cases the *not-isa* link specifies a nonsuperclass relationship but nevertheless requires a property to be inherited, namely the opposite of what would be typically inherited. This is questionable: If *is-a* is used to specify class relationships as well as inheritance, then *not-isa* should be used for explicitly ruling out certain class relationships (therefore stronger than no links at all) as well as not permitting any inheritance of properties of any sort. However, *not-isa* is usually used to convey the information that the negation of the property referred to by a class name is to be inherited. For instance, in the given example, the *Royal.elephant not-isa grey.thing* link is to be interpreted as: 'the class of Royal.elephants is not a subclass of grey.thing, *and also* the class of Royal.elephants is to inherit nongreyness so that its own subclasses and instances can inherit this property.' EIR drives a clear conceptual wedge between *is-a* and *not-isa* here: It is the exceptional class that conveys the missing, exceptional inheritable property, not the *not-isa* link. The exceptional class therefore allows the opposite or negation of what would be typically inherited to be inheritable in its own right. This would leave *not-isa* to perform the simple but important task of explicitly ruling out certain subclass-superclass relationships, without necessarily being involved in aspects of inheritance in any way. The role of EIR is to create an exceptional class that can be inserted into the structure so that the following, new path results:

$$Clyde \rightarrow Royal.elephant \rightarrow \Psi \not\rightarrow grey.thing$$

where Ψ is the exception class that is just like the *elephant* class, except that it is a subclass to a class (nongrey things) that stands for the opposite of what elephants typically are a subclass of (grey things).

Although the application of EIR to other inheritance problems is being examined, it is quite possible that its main application may well be in areas such as active object-oriented (o-o) databases (where production rule definitions interfaced to an o-o database can contain exceptions in the situation where an active database is not seen as a slave to an application, but as a peer that reacts to changes in the database state. In this sense, whereas an efficient monitoring of the active events and rules is essential, an explanation resolving conflict cases that may contain redundant information is needed), o-o expert systems (where diagnostic reasoning needs to take exceptions into account), and analogical reasoning (where mappings must be found in object-centered representations of base and target domains). The main contribution of EIR, however, is to offer a way of restructuring nonmonotonic inheritance structures so that conflict problems can be resolved that have otherwise proved intractable. Such restructuring is based on the intuitively obvious, but often forgotten distinctions between on the one hand, instantiation and inheritance logic (where the two have been confused by the use of *is-a* and *not-isa* to convey both styles of reasoning), and inheritable and noninheritable (i.e. acquired) properties, on the other.

A formal introduction to EIR, a number of examples and an exceptional inheritance algorithm has been defined and presented as well. In addition to exceptional information, I have treated ambiguity (in Chapter 6) as a problem of identifying a unique extension in the light of unrelated and conflicting classes. We defined ambiguity as a situation that contains several different properties that do not fit together and could be understood in more than one way. In real life, humans as intelligent entities perform comparisons by understanding the nature of each of the conflict classes separately, and jointly – that is, what are the properties of a class alone and what is the effect from other conflict classes on these properties? All such comparisons and eventual property preferences will be done by default. The successful default (preferred) among others will rise (according to its priority) to deal with such comparison. For instance, if we lose somebody in a jungle, we prefer to find that person first rather than eat our meal, even if we are hungry (notice there is no obvious relation between the hungry.person class and lost.somebody class). Or, we prefer a glass of water in a desert rather than a rest for 2 hours under a tree (although there is no obvious relation between water and resting). Also, we may prefer to consider a person who seems guilty of committing a crime innocent if we take into account their excellent record, so that we can reason that the event was accidental.

With this, I described Default Correlation (DC), an approach to inheritance-based reasoning with conflicts. This is done by defining the importance (weight) of the ambiguous property, or the level of importance of that property among other properties, regarding its typical classes. While the importance of the positive property regarding its typical class is measured, the importance of the negation of the same property regarding the other conflict class is measured as well. It is similar to say that we partition a default to its components (or levels). Representation language is presented and an algorithm based on what we termed *Default Correlation (DC)* to deal with this problem is also presented.

EIR was also used, as a base to assist defining *causal relations* between two domains. A causal relation was described as a relation between the top-level properties of two inheritance structures and not between the objects themselves. In other words, to say that an event causes another event is to say that an action causes a state (a property among other properties of an object) in a way that causes that event. The unexpected states of one action can be then explained using the exceptional inheritance classes (Ψ), which are located in the base of a causal relation. For example consider the following two statements;

Throwing 'my-glass' on 'my-floor' causes the breaking of 'my-glass'.

Throwing 'my-glass' on 'my-floor' causes the cracking of 'my-glass'. Those various states demonstrate the different situations, namely 'cracked' and 'broken', that an instance may be in via a specific action. Any knowledge representation system has a unique chance to explain one of these two situations. Regardless of the conditions that have a chance to influence causation, the point here is to show how EIR could represent the two situations, using the exceptional classes, effectively.

A formal introduction to inheritance-causal relation is presented in Chapter 7, and an exceptional inheritance-causal algorithm is proposed and defined as well.

Another application to EIR was demonstrated in a relatively complicated script-based story. EIR was applied to examine what an inheritance-based reasoner can do in script-based understanding and to give a causal explanation about why the story happened in a particular way that may or may not have followed the predefined scripts. For example, when people hear a story about entering a restaurant, they remember their past experiences about what *normally* happened in the restaurant: entered, found a seat, looked at the menu, and so on. All these actions have repeatedly taken place in such situations so that people are likely to follow such a script to fill gaps between isolated events. The problem arises when a story does not follow the script, (i.e., entered, spoke to the manager, left). Whatever the reason was, we need to explain why the new script did not follow the *typical* restaurant script. Through distinguishing implicit and explicit causality, the theory offers a more reasonable explanation of why the story goes in a particular way as opposed to following the normal course of the script.

The EIR's causal relation led to an examination of the potential use of EIR in a different reasoning domain, namely *analogical reasoning*. Because analogical reasoning uses well-defined relations in a well-based domain (the base) to assist in defining relations in an ill-based domain (the target), we have found that using EIR and causation can be of great help in defining the alternative classes in a target domain easily. This is done, in Chapter 8, by defining the shared top-level property of two structures. EIR assists in defining the alternative classes within each of the analogical inheritance structures, so that we will be able to create or suggest new classes that related to the same top-level property easily.

For example, Zaid's home is in Exeter and must travel to London. After discovering that his father's car is broken, Zaid is aware of the possibility of using the bus. Although Zaid did not use the bus before, he used the causal as well as the analogical reasoning in using the alternative to his father's old car; that is car, the instance of the class vehicle, is similar to the bus, instance of the same class, sharing the same property, carrying things for short or long distances, with some exceptions (Ψ).

In Chapter 8, a formal introduction to inheritance-analogical relation as well as number of examples were presented.

9.2 Outstanding problems

The framework of EIR is supported by its upward, off-path, skeptical approach. Having said that, an investigation of the different kinds of *not-isa* relations that are used within this approach is needed. To be precise, the translation of relational structures of the form *not-isa* as an opposite to *is-a* is not intuitively clear. For instance, we refer to Section

2.4 and the following paths:

$$Clyde \rightarrow Elephant \not\rightarrow Flying.things$$

and

$$Tweety \rightarrow Penguin \not\rightarrow Flying.things.$$

From this we can recognize that neither Clyde nor Tweety are flying.things. Clyde is recognized similarly to Tweety regarding one specific property. But the question that arises is: Are the semantics behind these two statements identical? Or, is it right for an inference rule to accept these statements as a ground base to two different applications? I believe that we need to integrate inductive reasoning (using EIR causal and analogical reasoning) and deductive reasoning (using EIR) in defining the similarity and the difference between two domains. Otherwise we will find ourselves using *not-isa* just as a nonsuperclass relation in two completely different domains.

Also in this book (Chapter 6) I suggested that ambiguity could be approached from different levels, (i.e, definition level, default level, and conflict classes level). Having reached ambiguity it is resolved using conflict classes to represent one explicit property as all other levels remain untackled. In other words, we need to explain what we follow in defining the conflict classes in the first place. For example, (using Figure 6.5(b) although we were able to define the conflict classes in the first place, we did not name their conflict property. Different definitions of this property will lead to different considerations. For more details regarding this point see Section 6.3.

Finally, we may end by saying that AI problems are very hard, and that is because they deal with the complex nature of human commonsense reasoning. The difficulty involved in formalizing commonsense reasoning is due to the nonmonotonic nature of this type of reasoning. Nonmonotonic reasoning is an area that many AI researchers are wary of. We are living in a dynamic world with new states of knowledge replacing old. Any practical nonmonotonic system that fails to take into account the fact that a new state may cancel out a previous one will be limited in its ability to capture the dynamic nature of our knowledge of the world. If a previous conclusion is shown to be false, not only must the earlier conclusion be retracted, but so must all the other conclusions drawn in the meantime from the earlier one. In other words, the problem with almost all existing AI systems lies in the difficulty of describing or formalizing human commonsense reasoning precisely or intuitively.

Such a formalization has been the goal of this book.

Index

Bibliography

Al-Asady, R. & Narayanan, A. (1993a). More notes on 'a clash of intuitions'. In *Proceedings of IJCAI-93*, pages 682–687. San Mateo, CA: Morgan Kaufmann.

Al-Asady, R. & Narayanan, A., editor (1996). *Inheritance Networks for Artificial Intelligence.* To be published.

Al-Asady, R. (1993b). A unified approach to inheritance with conflicts. In Ryan, K. & Sutcliffe, R., editor, *Proceedings of AICS-92 - The Fifth Irish Conference on Artificial Intelligence and Cognitive Science*, pages 291–294. London: Spinger-Verlag.

As-Sadr, M. B. (1987). *Our philosophy.* Ansariyan Pub., Qum, Iran.

Bacchus, F. (1989). A modest, but semantically well founded, inheritance reasoner. In *Proceedings of IJCAI-89*, pages 1104–1109.

Bobrow, D. and Winograd, T. (1985). An overview of KRL, a knowledge representation language. In Brachman, R. and Levesque, H., editors, *Reading in knowledge representation*, pages 263–285.

Boutilier, C. (1989). A semantic approach to stable inheritance reasoning. In *Proceedings of IJCAI-89*, pages 1134–1139.

Brachman, R. (1983). What isa is and isn't: an analysis of taxonomic links in semantic networks. *IEEE Computer*, 16(10):30–36.

Brewka, G. (1987). The logic of inheritance in frame systems. In *Proceedings of IJCAI-87*, pages 483–488.

Brewka, G. (1989). Preferred subtheories: An extended logical framework for default reasoning. *Proceedings of IJCAI-89*, pages 1043–1048.

Brewka, G. (1991). *Nonmonotonic Reasoning: Logical Foundations of Commonsence.* Cambridge: Cambridge University Press.

Burstein, M. (1986). Concept formation by incremental analogical reasoning and debugging. In Carbonell, J. and Michalski, R.S. & Mitchell, T., editors, *Machine learning: An artificial intelligence approach*, volume 2, pages 351–369. Tioga, Palo Alto, CA.

Carbonell, J. (1986). Derivational analogy: A theory of reconstructive problem solving and expertise acquisition. In Carbonell, J. and Michalski, R.S. & Mitchell, T., editors, *Machine learning: An artificial intelligence approach*, volume 2, pages 371–392. Tioga, Palo Alto, CA.

Carbonell, J. and Michalski, R.S. & Mitchell, T. (1983a). Learning by analogy: Formulating and generalising plans from past experience. In Carbonell, J. and Michalski, R.S. & Mitchell, T., editors, *Machine learning: An artificial intelligence approach*, volume 1, pages 137–161. Tioga, Palo Alto, CA.

Carbonell, J. and Michalski, R.S. & Mitchell, T. (1983b). An overview of machine learning. In Carbonell, J. and Michalski, R.S. & Mitchell, T., editors, *Machine learning: An artificial intelligence approach*, volume 1, pages 3–24. Tioga, Palo Alto, CA.

Cox, P.T. & Pietrzykowski, T. (1986). Causes and events: Their computation and applications. In *Proceedings of Eight int. Conf. on Autom. Deduction. Lecture Notes in Computer Science 230*, pages 608–621.

Davis, E. (1990). *Representations of Commonsense Knowledge.* Morgan Kaufmann, San Mateo, CA.

Delgrande, J. (1987). A first order logic for prototypical properties. *Artificial Intelligence*, 33:105–130.

Delgrande, J. (1988). An approach to default reasoning based on a first-order conditional logic: Revised report. *Artificial Intelligence*, 36:63–90.

Ellman, T. (1989). Explanation-based learning: A survey of programs and perspectives. *ACM Computing Surveys*, 21(2):163–221.

Etherington, D. (1987a). Formalizing nonmonotonic reasoning systems. *Artificial Intelligence*, 31:41–85.

Etherington, D. (1987b). More on inheritance hierarchies with exceptions: Default theories and inferential distance. In *Proceedings of AAAI-87*, pages 352–357.

Etherington, D. (1987c). A semantic for default logic. In *Proceedings of IJCAI-87*, pages 495–498.

Etherington, D. (1988). *Reasoning with Incomplete Information: Investigation of Nonmonotonic Reasoning.* Morgan Kaufmann, San Mateo, CA.

Etherington, D. (1994). Brief comments on 'formalizing nonmonotonic reasoning systems'. In Al-Asady, R. & Narayanan, A., editor, *Inheritance Networks for Artificial Intelligence.* To be published.

Etherington, D. and Reiter, R. (1983). On inheritance hierarchies with exceptions. In *Proceedings of AAAI-83*, pages 104–108.

Evans, T. (1968). A program for the solution of geometric analogy intelligence test questions. In Minsky, M., editor, *Semantic information processing*, pages 271–353. MIT Press, Cambridge.

Fahlman, S. (1979). *NETL: A System for Representing and using Real-World Knowledge.* MIT Press, Cambridge, MA.

Falkenhainer, B. and Forbus, K. & Gentner, D. (1989). The structure-mapping engine: Algorithm and examples. *Artificial Intelligence*, 41:1–63.

Forbus, K.D. & Gentner, D. (1986). Learning physical domains: Toward a theoretical framework. In Carbonell, J. and Michalski, R.S. & Mitchell, T., editors, *Machine learning: An artificial intelligence approach*, volume 2, pages 311–348. Tioga, Palo Alto, CA.

Galton, A. (1991). A critique of Yoav Shoham's theory of causal reasoning. In *Proceedings of AAAI-91*, pages 355–359.

Gan, H. & Al-Asady, R. (1993). A new formalization for script-based story understanding. In *Proceedings of the 6th conference on AICS-93, Queen's University, UK.*, pages 175–184. ISSN.

Geffner, H. & Verma, T. (1989). Inheritance = chaining + defeat. In *Proceedings of Methodologies for Intelligent Systems*, pages 411–418.

Gentner, D. (1982). Structure mapping: A theoretical framework for analogy and similarity. In *Proceedings of the Fourth Conference of the Cognitive Science Society*, pages 13–15.

Gentner, D. (1983). Structure-mapping: A theoretical framework for analogy. *Cognitive Science*, 7:55–70.

Ginsberg, M. (1986). Counterfactuals. *Artificial Intelligence*, 30(1):35–81.

Ginsberg, M. (1988). Multivalued logics: A uniform approach to reasoning in artificial intelligence. *Computer Intelligence*, 4(3):265–316.

Ginsberg, M. (1991). A local formalization of inheritance: Preliminary report. In *Nonmonotonic Workshop-Stanford University.*

Hall, R. (1989). Computational approach to analogical reasoning: A comparative analysis. *Artificial Intelligence*, 39(1):39–120.

Haugh, B. (1989). *Nonmonotonic formalisms for commonsence temporal-causal reasoning.* PhD thesis, U.M.I., Dept. of Philosophy.

Hayes, P. (1977). In defence of logic. In *Proceedings of IJCAI-77*, volume 1, pages 559–565.

Hayes, P. J. (1979). the logic of frames. In Webber, B. L. and Nilsson, N. J., editors, *Readings in artificial intelligence*, pages 451–458. Tioga, CA.

Horty, J., Thomason, R., and Touretzky, D. (1987). A skeptical theory of inheritance in nonmonotonic semantic networks. In *Proceedings of the Sixth National Conference on Artificial Intelligence*, pages 385–362.

Horty, J. & Thomason, R. (1988). Mixing strict and defeasible inheritance. In *Proceedings of AAAI-88*, pages 427–432.

Israel, D. (1987). What's wrong with non-monotonic logic? In Ginsberg, M., editor, *Reading in nonmonotonic reasoning*, pages 53–55. Morgan Kafmann, CA.

Kedar-Cabelli, S. (1985). Purpose-directed analogy. In *Proceedings of 7th Annual Conference of the Cognitive Science Society*, pages 150–159. Irvine, CA.

Kedar-Cabelli, S. (1988). Analogy-from a unified perspective. In Helman, D., editor, *Analogical reasoning-perspectives of artificial intelligence, cognitive science and philosophy*, volume 197, pages 65–103. Kluwer academic publishers.

Kedar-Cabelli, S. (1990). Toward a computational model of purpose-directed analogy. In shaulik, J. and Dietterich, T., editors, *Reading in machine learning*, pages 647–656. San Mateo, CA: Morgan Kaufmann.

Kling, R. E. (1971). A paradigm for reasoning by analogy. *Artificial Intelligence*, 2(2):147–178.

Kodratoff, Y. (1988). *Introduction to Machine Learning*. London: Pitman.

Konolige, K. (1988a). Hierarchic autoepistemic theories for nonmonotonic reasoning. In *Proceedings of AAAI-88*, pages 439–443. Morgan Kafmann.

Konolige, K. (1988b). On the relation between default and autoepistemic logic. *Artificial Intelligence*, 35(3):343–382.

Konolige, K. (1989). On the relation between autoepistemic logic and circumscription- preliminary report. In *Proceedings of IJCAI-89*, pages 1213–1218.

Krishnaprasad, T. & Kifer, M. (1989). An evidence-based framework for a theory of inheritance. In *Proceedings of IJCAI-89*, volume 2, pages 1093–1098.

Krishnaprasad, T. and Kifer, M. & Warren, D. (1989). On the declarative semantics of inheritance networks. In *Proceedings of IJCAI 89*, pages 1099–1103.

Lewis, D. (1973). Causation. *Journal of Philosophy*, 70:556–567.

Lifschitz, V. (1986a). On the satisfiability of circumscription. *Artificial Intelligence*, 28:17–27.

Lifschitz, V. (1986b). Pointwise circumscription. In *Proceedings of AAAI-86*, pages 406–410.

McCarthy, J. (1977). Epistemological problems of artificial intelligence. In *Proceedings of IJCAI-77*, pages 1038–1044.

McCarthy, J. (1980). Circumscription-a form of non-monotonic reasoning. *Artificial Intelligence*, 13:27–39.

McCarthy, J. (1986). Application of circumscription to formalizing commonsense knowledge. *Artificial Intelligence*, 28:89–116.

McDermott, D. (1982). Nonmonotonic logic II: Nonmonotonic model theories. *JACM*, 29(1):33–57.

McDermott, D. and Doyle, J. (1980). Nonmonotonic logic I. *Artificial intelligence*, 13:41–72.

Minsky, M. (1975). A framework for representing knowledge. In Winston, P., editor, *The Psychology of Computer Vision*, pages 211–277. MIT Press, Cambridge.

Moore, R. C. (1984). Possible-worlds semantic for autoepistemic logic. In Ginsberg, M., editor, *Reading in nonmonotonic reasoning*, pages 137–142. San Mateo, CA: Morgan Kafmann.

Moore, R. C. (1985). Semantic considerations on nonmonotonic logic. *Artificial Intelligence*, 25:75–94.

Neufeld, E. (1991). Notes on 'A clash of intuitions'. *Artificial Intelligence*, 48(2):225–240.

Neufeld, E. and Poole, D. (1990). Probabilistic semantic and defaults. In Scachter, R., Levitt, T., Lemmer, J., and Kanal, L., editors, *Uncertainty in artificial intelligence 4*, pages 275–282. Elsevier, The Netherlands.

Padgham, L. (1989). Negative reasoning using inheritance. In *Proceedings in IJCAI 89*, volume 2, pages 1086–1092.

Padgham, L. (1991). A lattice-based model for inheritance reasoning. In Lenzerini, M. and Nardi, D. & Simi, M., editors, *Inheritance hierarchies in knowledge representation and programming languages*, pages 51–69. John Wiley.

Padgham, L. (1992). Defeasible inheritance: A lattice-based approach. *Computer Math. Applic.*, 23(6-9):527–541.

Patel-Schneider, P. (1991). What's inheritance got to do with knowledge representation. In Lenzerini, M. and Nardi, D. & Simi, M., editors, *Inheritance hierarchies in knowledge representation and Programming Languages, 1-11.* John Wiley & Sons.

Pearl, J. (1988). Embracing causality in default reasoning. *Artificial Intelligence*, 35:259–271.

Poole, D. (1988). A logical framework for default reasoning. *Artificial Intelligence*, 36:27–47.

Poole, D. (1991). The effect of knowledge on belief: conditioning, specificty and the lottery paradox in default reasoning. *Artificial Intelligence*, 49:281–307.

Quillian, M. R. (1968). Semantic memory. In Minsky, M., editor, *Semantic information processing*, pages 227–270.

Reiter, R. & Criscuolo, G. (1981). On interacting defaults. In Ginsberg, M., editor, *Reading in nonmonotonic reasoning*, pages 270–276. San Mateo, CA: Morgan Kafmann.

Reiter, R. (1980). A logic for default reasoning. *Artificial Intelligence*, 13:81–132.

Reiter, R. (1987). A theory of diagnosis from first principles. In Ginsberg.M, editor, *Reading in nonmonotonic reasoning*, pages 352–371. Morgan Kafmann.

Robert, B. & Goldstein, P. (1977). *The FRL manual. AI Memo No. 409.* Cambridge University Press.

Rumelhart, D.E. & Norman, D. (1981). Analogical processes in learning. In Anderson, J., editor, *Cognitive skills and their acquisition*, pages 45–70. Hillsdale,N.J: Erlbaum.

Sandewall, E. (1986). Nonmonotonic inference rules for multiple inheritance with exceptions. In *Proceedings of IEEE-86*, volume 74, pages 1345–1353.

Schank, R. (1975). *Conceptual Information Processing.* North-Holland, New York.

Schank, R. (1986). *Explanation Patterns, Understanding Mechanically and Creatively.* Lawrence Erlbaum.

Schank, R. and Abelson, R. (1977). *Scripts, Plans, Goals and Understanding: An Inquiry into Human Knowledge Structures.* Lawrence Erlbaum Associates.

Schubert, L. K. (1976). Extending the expressive power of semantic networks. *Artificial Intelligence*, 7:163–198.

Selman, B. & Levesque, H. (1989). The complexity of path-based defeasible inheritance. In *Proceedings of IJCAI-89*, pages 1140–1145.

Selman, B. (1990). *Tractable default reasoning.* PhD thesis, University of Toronto.

Shastri, L. (1989). Default reasoning in semantic networks: A formalization of recognition and inheritance. *Artificial Intelligence*, 39:283–355.

Shoham, Y. (1988). *Reasoning about change: Time and causation from the standpoint of artificial intelligence.* MIT Press.

Sowa, J. (1992). *Principles of semantic networks.* San Mateo, CA: Morgan Kaufmann.

Stein, L. (1989). Skeptical inheritance: computing the intersection of credulous extensions. In *Proceedings IJCAI 89*, pages 1153–1158.

Stein, L. (1991). Computing skeptical inheritance. In Lenzerini, M. and Nardi, D. & Simi, M., editors, *Inheritance hierarchies in knowledge representation and programming languages*, pages 69–81. John Wiley & Sons.

Suppes, P. (1970). *A probabilistic Theory of Causation.* North Holland.

Thayse, A., editor (1988). *From standard logic to logic programming.* John Wiley & sons, New York.

Thomason, R. & Horty, J. (1989). Logics for inheritance theory. In *Proceedings of the 2nd International Workshop on Nonmonotonic Reasoning*, pages 220–237.

Thomason, R., Horty, J., and Touretzky, D. (1987). A calculus for inheritance in monotonic semantic nets. In Ras, Z. & Zemankova, M., editor, *Proceedings of the Second International Symposium on Methodologies for Intelligent Systems*, pages 280–287.

Touretzky, D. (1984). Implicit ordering of defaults in inheritance systems. *Proceedings of AAAI-84*, pages 332–325.

Touretzky, D. (1986). *The mathematics of inheritance systems.* London: Pitman.

Touretzky, D., Horty, J., and Thomason, R. (1987). A clash of intuitions: the current state of nonmonotonic inheritance systems. *Proceedings of IJCAI-87*, pages 476–482.

Winston, P. (1975). Learning structure description from examples. In Winston, P., editor, *Psychology of computer vision*. McGraw-Hill.

Winston, P. (1980). Learning and reasoning by analogy. *Commun. ACM*, 23(12):689–703.

Winston, P. (1984). *Artificial Intelligence*. Addison-Wesley, 2nd edition.

Woods, W. (1975). What in a link: Foundations for semantic networks. In Bobrow, D.G. & Collins, A., editor, *Representation and understanding: Studies in cognitive science*, pages 35–82. Academic press, New York.